Cambridge IGCSE®
Biology

Revision Guide

Ian J. Burton

CAMBRIDGE
UNIVERSITY PRESS

University Printing House, Cambridge CB2 8BS, United Kingdom

One Liberty Plaza, 20th Floor, New York, NY 10006, USA

477 Williamstown Road, Port Melbourne, VIC 3207, Australia

4843/24, 2nd Floor, Ansari Road, Daryaganj, Delhi – 110002, India

79 Anson Road, #06–04/06, Singapore 079906

Cambridge University Press is part of the University of Cambridge.

It furthers the University's mission by disseminating knowledge in the pursuit of education, learning and research at the highest international levels of excellence.

www.cambridge.org
Information on this title: education.cambridge.org

First published 2016

20 19 18 17 16 15 14 13 12 11 10 9 8 7 6 5 4 3 2

Printed in India by Repro India Ltd

A catalogue record for this publication is available from the British Library

ISBN 978-1-107-614499 Paperback

Table of Contents

Preface

Revision tips for IGCSE biology students

Understanding IGCSE Biology is not usually a problem, but committing facts to memory can often be a major obstacle to success. Many students are at a loss to know exactly how to set about what seems to them to be a task of immense proportions. I offer the following method, one that I devised myself when, as a student, I was faced with the same problem. It has the advantage, if followed carefully, of improving one's factual knowledge as a result of time spent.

This revision guide is full of important terms and phrases. The method that I offer for learning it is as follows:

1. Take a sheet of file paper and divide it with a vertical line such that three quarters of the sheet is on the left of the line.

2. Read a page of the revision text and, each time you come to an important word or phrase, think up your own simple question to which that word or phrase is the answer.

3. Write these simple questions on the left hand side of your sheet of file paper, leaving a space between each, and number them. Continue on further sheets of paper if necessary.

4. If there is a diagram in the text, then draw a quick sketch of the diagram on the left-hand side of your sheet with numbered label lines vertically above each other extended towards the right-hand side of your sheet.

5. When you have reached the bottom of the page of text, close the book and see how many of the answers you can write down on the right hand side of your sheet. When you have attempted all answers, check them against the text. You will probably be surprised at how well you do, but since you wrote the questions, carefully phrased around the required answer, perhaps it is not so surprising after all.

6. Continue until you have a list of questions and answers to the section you are trying to learn.

7. Take a second sheet of paper (folded if writing would otherwise show through it) and use this to cover the answers. Test yourself again, writing your answers on the folded sheet, and continue this until you are able to score over 80%. (You can, of course, set your own target. Some will not be content until they can score 100%.)

8. File away your Question/Answer sheet for further revision at a later date.

9. Continue this process systematically until you have, effectively, a full set of revision notes for later use.

10. In the last few weeks before the examination, it is better to revise by reading the text of this book carefully, section at a time. Concentrate on every sentence, making sure you understand what you have read. It is so easy to get to the bottom of a page in a book and realise that your mind was elsewhere as you were reading it and, as a result, nothing registered at all. If that happens, be honest with yourself. Go back to the top of the page and start again.

11. In the last few days before the examination, your Question/Answer sheets should now prove invaluable for last-minute consolidation of your facts.

It cannot be stressed too strongly that examination results depend on knowledge. It is important that you have a very good grasp of simple knowledge to do well, and interpretation questions rely heavily on a sound knowledge of the subject matter.

The advantage of this revision method is based so firmly on the student phrasing the questions to which he or she will already know the answer that it would defeat the object if more than a short example of the technique were given. The success of the method relies only on the student following the technique carefully. It does work, but you must be prepared to spend the necessary time. You may even enjoy the experience!

Example of a Revision Sheet, based on information in this revision guide:

1. What word is used for organisms containing only one cell?   unicellular

2. Give an example of a one-celled organism.   a bacterium

3. What word is used for organisms made of many cells?   multicellular

4. What structure controls the passage of substances into and out of a cell?   cell membrane

5. In what state must all chemicals be before they can enter or leave a cell?   in solution

6. What is the jelly-like substance where chemical reactions occur in a cell?   cytoplasm

7. What is the correct term for the chemical reactions in a cell?   metabolic reactions

8. Whereabouts in a cell are chromosomes found?   the nucleus

9. What do chromosomes contain?   genes

10. Of what chemical are chromosomes made?   DNA

11. What makes up protoplasm?   cytoplasm +nucleus

12. What is the space in the centre of a plant cell?   vacuole

13. What does the space in the centre of a plant cell contain?   cell sap

14. What is the name of the box in which a plant cell is contained?   cell wall

15. What chemical is this box made of?   cellulose

16. Name the green structures in photosynthesising cells.   chloroplasts

17. What pigment do they contain?   chlorophyll

Here is how you can test yourself on the labels to a diagram.

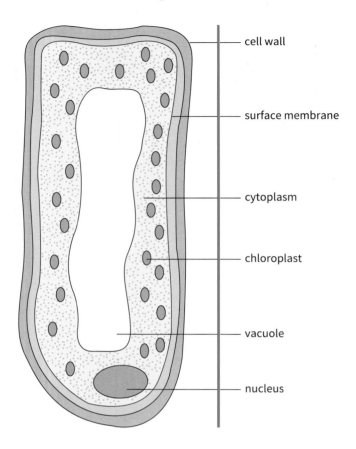

- cell wall
- surface membrane
- cytoplasm
- chloroplast
- vacuole
- nucleus

As the examination approaches and a greater amount of time is spent on revision, it is usually more productive to set aside a certain time each day for revision. Do not allow yourself to be persuaded to do anything else during that time.

Work **on your own** with **no distractions around you**. Some people say they can work better listening to music. If that really is so in your case, then keep the music quiet and, at least, it may shut out other distractions!

You may find it helpful to make a calendar by dividing a piece of paper into a space for each day during your revision period before the examination. Then you can divide the syllabus into the same number of parts as there are days for revision and enter one such part per day on your calendar. In this way you will know exactly what you are going to revise on each day. Your day's revision will not be complete until you have revised everything on your calendar for that day.

People vary as to how long they can work at a stretch. It is important to have a break from time to time (again, preferably, the same time each day). When you stop, set yourself a time to resume your revision **and stick to it**.

It would indeed be a pity if, armed with a sound factual knowledge, you then failed to use that knowledge effectively in the examination. You may find the following advice useful:

Teacher's tips for answering examination questions

1. It can **never** be said often enough: **read the question**. You must answer the question asked, not any other. It is probable that, on first reading, you feel worried that you can't answer it. You may think you have never learnt that particular topic. But, wait a moment, then read it again and you will often be surprised how much more sense it makes the second time!

2. Do not omit facts simply because you consider them too obvious to mention. They will often be credited.

3. The space provided for your answer on the question paper is a guide to how long your answer should be. Don't waste time or space writing irrelevant material.

4. Make sure that you do not contradict yourself. You are unlikely to get credit for a correct statement when you have also stated the contradiction.

5. If a question involves the interpretation of a graph, try to include some numerical information read from the graph and, if they are available, remember to include the appropriate units.

Finally, good luck with your revision. This method can work. I know, because it did so for me!

Ian J. Burton

How to use this book

Learning outcomes

By the end of this chapter you should understand:

- ☐ The characteristics of living organisms
- ☐ How to use the binomial system for naming organisms
- ☐ How living organisms are classified
- ☐ The characteristics of some vertebrates
- ☐ The characteristics of some invertebrates
- ☐ Viruses, prokaryotes (bacteria), protoctists and fungi
- ☐ The construction and use of a dichotomous key

Learning outcomes – set the scene of each chapter, help with navigation through the book and give a reminder of what's important about each topic.

Supplement material – indicated by a bold vertical line. This is for students who are taking the Extended syllabus covering the Core and Supplement content.

Terms – words in bold indicate important terms and definitions that are explained clearly in each topic.

Tips – quick suggestions to remind you about key facts and highlight important points.

Glossary terms – terms in green can also be found in the Glossary

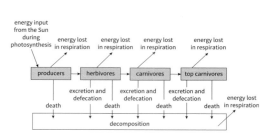

Figure 20.1 Energy flow in an ecosystem

- Each organism unlocks some of this energy to **use** for various processes within its body, for example, making new cells and the large organic molecules within them (during growth), muscular contraction (and movement), generating electrical impulses in the nervous system and raising body temperature. The chemical reaction that unlocks the chemical energy for conversion into other forms is respiration.
- Energy is used up in most of these processes. Only in the form of heat from an organism's body is it released to the environment outside the food chain. This includes energy from the respiration of bacteria and fungi that eventually decay dead organisms.

 TIP Avoid saying that energy is 'needed' for respiration (it is **released** by respiration).

Much of the energy is still present in the faeces and some in the nitrogenous waste of animals. This energy is available to decomposers. Not all herbivores are eaten, thus the amount of energy left within herbivores to be passed on to carnivores is small – 20% (only 2% of the original amount in the producer).

For this reason, food chains are limited in length, as there is insufficient energy remaining to sustain a succession of carnivores. **Five** trophic levels are usually the limit for a food chain (Figure 20.1).

The longer the food chain, the less the energy available to the top carnivore at the end of the chain. Short food chains are therefore much more energy efficient than

long ones. In order to supply enough energy in food to maintain an ever increasing world population, it must be realised that far less energy is lost when man eats green plants than when crop plants are fed to animals, which are then eaten by a human.

In any one habitat, such as a pond or mangrove swamp, there will be many organisms living together. In some way they will all be interconnected by way of different food chains.

A **network of interconnected food chains** is known as a food web (Figure 20.2).

While all food chains (and thus all food webs) begin with a producer, food webs may begin with several different species of producer.

A **producer** is an organism that makes its own organic nutrients, usually using energy from sunlight, through photosynthesis.

In a food chain or food web, producers are eaten by consumers.

A consumer in a food chain is an organism that gets its energy by feeding on other organisms.

An animal that gets its energy by eating plants is an herbivore (or primary consumer).

An animal that gets its energy by eating other animals is a carnivore (or secondary consumer – the consumer that feeds on the secondary consumer is a tertiary consumer – and so on). Thus all consumers above the level of herbivore, that is, all meat eaters, are carnivores.

When all organisms in a food chain or web die they are decomposed largely by bacteria and fungi.

161

PRACTICAL

A demonstration of the effects of a lack of nitrates and of magnesium ions in a growing plant

Apparatus: 2 small cuttings or seedlings
(e.g. sorghum)
2 containers
Cotton wool
Black paper or black polythene
Culture solutions

Method:

Two seedlings or small cuttings, with the same number of leaves, are selected from a quick-growing plant and held in the top of two containers (A and B) using cotton wool as shown in Figure 6.10.

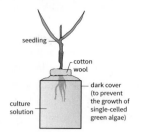

Figure 6.10 Experiment to show effects of a lack of nitrates and of magnesium ions in a growing plant

→

Practical skills – reinforce your practical knowledge and skills with clear explanations and diagrams.

Make sure you turn the page when you see this arrow in the corner!

Progress check 13.3

1   It is easier to see the outline of an object in dim light by looking to the side of it. Can you explain why this is?

2   Why do you think that a 'round-arm', punch aimed at the side of a boxer's face is often more successful than one aimed straight at him?

Progress check questions – check your own knowledge and see how well you're getting on by answering regular questions.

Worked example

Describe how the structure of a cell membrane is adapted to the process of active uptake.

Cell membranes contain protein molecules called 'channel proteins'. These molecules fit loosely together leaving minute channels between them extending from the outside of the cell to the cell cytoplasm. Suggest how these channel proteins may play a part is the processes of diffusion and osmosis.

Answer

The question calls first for a realisation that there is a special structural feature of the cell membrane making it able to undergo active uptake. That feature is the presence of carrier proteins. Since they have to work against a concentration gradient (there may be a lower concentration of the chemical to be absorbed outside than inside the cell) then energy must be used. All energy within a cell is initially released by respiration and, in this case,

the energy is used to move the carrier proteins. First the carrier protein opens to the outside and allows the molecule to be absorbed to attach (bind) to the protein. Only one type of molecule will bind as the site is not suitable for any other molecule. The protein then changes shape again (again using energy) – closing the outer opening, and opening into the cytoplasm of the cell. The molecule is released into the cell cytoplasm, then changes shape again, closing to the inside and opening to the outside ready to bind with another molecule to be absorbed. (Note that the complete cycle of carrier protein movements is described as a continuous process.)

Diffusion requires pores in the membrane before molecules can enter. The channel proteins would provide the pores. However, the size of the pores might prevent larger molecules from entering. Water is a small molecule and thus could enter by osmosis, and the pores may be too small to allow larger molecules to enter – making the membrane semi-permeable.

Worked examples – a step by step approach to answering questions, guiding you through from start to finish.

Chapter summary

■ You have learnt how cells are involved in the processes of diffusion, osmosis and active transport.

■ You have learnt how these processes are important to living structures.

■ You have learnt how to demonstrate these processes experimentally.

■ You have also learnt about the factors that affect them.

Chapter summary – at the end of each chapter so you can check off topics as you revise them.

Exam-style questions

1   Describe how different substances in a leaf move by diffusion during a 24-hour period. [6]

2   a   Figure 3.13 shows a piece of partially permeable tubing, tightly tied at each end, and containing a concentrated sugar solution that is coloured with blue dye. It has been placed in a beaker of pure water:

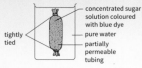

tightly tied

concentrated sugar solution coloured with blue dye

pure water

partially permeable tubing

Figure 3.13

Describe what will happen to the tubing and its contents over the next 20 minutes. [3]

b   After 20 minutes, apart from what happens to the tubing, the water in the beaker has turned blue. With reference to diffusion and osmosis, explain the results of this experiment. [7]

3   a   Explain how a plant root absorbs from the soil:

i) water [6]

ii) essential mineral ions that are in very short supply. [4]

b   Suggest why a plant may have great difficulty in absorbing essential mineral ions that are in very short supply in a water-logged soil. [4]

Exam-style questions – prepare for examinations by completing the Exam-style questions and checking your answers, which are provided at the back of the book.

Acknowledgements

Every effort has been made to trace the owners of copyright material included in this book. The publishers would be grateful for any omissions brought to their notice for acknowledgement in future editions of the book. The authors and publishers acknowledge the following sources of copyright material and are grateful for the permissions granted.

Cover image Frans Lanting, Mint Images/SPL; P.7c Alhovik/Shutterstock; P.11 AS Food studio/Shutterstock; P.15b Jose Luis Calvo/Shutterstock; P.17t Keith R. Porter/Science Photo Library; P.17b clusterx/Alamy; P.18l Power and Syred/Science Photo Library; P.19 Dr. Robert Calentine, Visuals Unlimited/Science Photo Library; P.20r Dr David Furness, Keele University/Science Photo Library; P.50r Dr Keith Wheeler/Science Photo Library; P.83b Dr. Fred Hossler, Visuals Unlimited/Science Photo Library; P.84 Eye of Science/Science Photo Library; P.126tl Dr Jeremy Burgess/Science Photo Library; P.126tr Susumu Nihhinaga/Science Photo Library; P.156t AlexSmith/Shutterstock; P.156b small1/Shutterstock.

SPL = Science Photo Library

Classification

1.01 Characteristics of living organisms

All living **organisms** possess the 'characteristics of life'. The one group of organisms that does not show all the characteristics of life is the **viruses**. Thus they are considered to be on the border between living and non-living.

All truly living organisms display the following characteristics:

- **Movement.** This may be the movement of a part of an organism in relation to the rest of its body (such as the movement of an arm or of a shoot tip), or it may involve the movement of the whole organism from one place to another – when it is called locomotion. It commonly involves the contraction of muscles (as in the arm) or cells growing at different rates (as in the shoot tip).

  It is thus defined as **an action by an organism (or part of an organism) causing a change of position or place**.

- **Respiration.** This is a chemical reaction that takes place in living cells. It involves the breakdown of large, nutrient organic molecules (usually **carbohydrates**, such as glucose) to release (not to 'make', 'manufacture' or 'produce') the energy contained within the molecule. The glucose molecule contains energy in the form of chemical energy, which is converted into other forms for use in doing work – such as electrical energy in nerve impulses.

  Respiration is defined **as the chemical reactions in cells that break down nutrient molecules and release energy (for metabolism)**.

TIP
Breathing and respiration are not the same thing. When we take air into and expel air from our lungs, we are breathing. This process is to supply oxygen to the blood that takes it to the cells where respiration occurs.

- **Sensitivity.** This is the ability to detect and respond to changes in the environment (known as **stimuli**). The stimuli may be from the internal environment – for example, the effect of **hormones** on a cell or tissue, or from the external environment – for example, light. The internal environment is a term that refers to the conditions inside an organism. Sensitivity is also the ability to detect or sense stimuli in the internal or external environment and to make appropriate responses.

- **Growth.** It is customary for organisms to start life small in size and gradually become larger with time. Some organisms grow to a certain size then stop, while others grows continuously throughout their lives. Growth is defined as **a permanent increase in size**.

  Growth involves an increase in dry mass by an increase in cell number or cell size or both. Dry mass is the mass of all the components within an object except any water present.

- **Reproduction.** In order to maintain (or increase) their numbers, all organisms have the ability to make more of the same kind.

- **Excretion.** This is the removal from organisms of toxic materials and substances in excess of requirements.

  The material removed includes the waste products of **metabolism** – chemical reactions in cells including respiration.

> **TIP**
> Remember that excretion does NOT include the removal of undigested waste from the intestines since it has never taken part is a chemical reaction within the body's cells.

- **Nutrition.** In order to provide the raw materials and the energy for all the other characteristics of life listed previously, organisms must take in energy-containing materials that are required for growth and development.

  Nutrition is thus defined as **the taking in of materials for energy, growth and development**.

  Plants require light, carbon dioxide, water and ions; animals need organic compounds and ions and usually need water.

> **TIP**
> The first letters of each of the characteristics together spell the name of 'MRS GREN' – a lady well known to students trying to remember the characteristics of living organisms!

1.02 The concept and use of a system of classification

The living universe comprises well over 10 million different types of organism, which are sorted into groups based on common features. This is called **classification** (or **taxonomy**). Those organisms that share many similar features are placed in the same group. Those that share few features are placed in separate groups. The number of shared features between different groups gives an indication of how closely related the groups may be.

The largest groups are called **kingdoms**, of which there are five:

- Prokaryote (Bacteria)
- Protoctist
- Fungus
- Plant
- Animal.

Each kingdom is divided into sub-groups and each sub-group is divided into smaller groups. The last two groups in this succession are the **genus** and finally, the **species**. (The plural of genus is genera.)

A **species** is defined as **a group of organisms that can interbreed (reproduce) and produce fertile offspring**. A species is therefore said to be 'reproductively isolated'.

Organisms within a species are not identical and the differences between them are called **variations**.

The binomial system of classification

'bi' = two and 'nomial' from the Latin *nomen* = name

All living organisms are usually known by the **binomial system**, an internationally agreed system using two names.

These two names indicate the **genus** and the **species** to which the organism belongs.

The genus is always written with an upper-case first letter and the species is written with all lower-case letters. Both names are always underlined when hand-written and appear in italics in print. Both names often have a Latin or Greek origin. Thus, the lion is

<u>Panthera leo</u> (hand-written) and
Panthera leo (in print).

The binomial system is useful because:

- Sometimes, different species in different parts of the world share the same name. When different countries work together on schemes to conserve endangered species, it is vital that they are all considering the same organisms (e.g. there are three different species of arthropod all called 'Daddy Long-legs' in different parts of the world.)

- The same species may have different names in different languages.

- The common name may be misleading (a jellyfish is not a fish).

- All organisms placed in the same genus will share a set of features common only to that group. Knowing the genus, even without actually seeing the organism, therefore tells the biologist a great deal about organisms and about their evolutionary history and relationships (i.e. how recently they separated from one another as they have evolved).

For many years, the classification of organisms was based on studies of their morphology, that is, their outward appearance, for example, the number and type of limbs, or the shape of the flowers produced by a plant. It may include internal morphological features, such as the skeleton (useful when classifying fossils, for example). These studies were also supported by consideration of shared anatomical features, that is, internal features visible as a result of dissection of organisms.

RNA and DNA sequencing

The sequence of chemical bases in the DNA and RNA molecules found in different organisms gives a very accurate indication of how closely related those organisms are. Mutations are constantly changing this sequence and those changes are handed on to the next generation. (See Chapter 18.)

The sequence of bases in the DNA molecule determines the sequence of amino acids in the proteins made by the organism. Thus, a mutation in an organism's DNA leads to a change in its protein structure. The longer ago the two different organisms separated from a common ancestor, the larger the number of mutations will have occurred, and the greater the differences in the sequence of bases there will be in these organisms' DNA and RNA. This, in turn, leads to a greater difference in the amino acid sequence in their proteins.

Data from the analysis of DNA/RNA base sequences is now so accurate that we are able to identify human beings in the same family.

Progress check 1.1

1   Find out what you can about a DNA molecule. How many bases are there?

2   Make sure you know what each of the letters in 'Mrs Gren' stand for.

3   What describes respiration?

  A   breathing in oxygen

  B   breathing out carbon dioxide

  C   releasing energy from nutrient molecules

  D   using energy to construct nutrient molecules

1.03 Features of organisms

All living organisms share the possession of a cellular structure, that is, they are all made up of one or more living units called cells.

Cells include the following features:

- Cytoplasm – a jelly-like substance that contains smaller structures (organelles) and in which all the metabolic chemical reactions occur.

- DNA – the chemical that forms the genes of the cell that are responsible for the nature of the proteins made within the cell and also for handing on this information to future generations.

- Cell membrane – the living, selectively permeable structure that encloses the cell contents and is responsible for the entry of substances into and exit of substances from the cell.

Two of the most familiar kingdoms in the living universe are the animals and the plants. The distinguishing features of these two kingdoms are as follows:

Animals

- Animals take in (ingest) and use organic materials from other living organisms as their source of energy for growth and development.

- Animals are able to move from one place to another (movement known as locomotion). (Sponges are exceptions to this as they are animals that remain fixed to the surface on which they live.)

- Sexual reproduction – animals reproduce using specialized reproductive cells (gametes). The male gamete is the sperm and the female gamete the egg cell (or ovum). Few animals reproduce by asexual reproduction.

- Most animals have diploid nuclei. That is to say that each nucleus has two full sets of genetic material contained in matching chromosomes. Only the X and Y chromosomes (the sex chromosomes) do not exactly match.

- There is no rigid cell wall surrounding the cell membrane.

Plants

- Plants manufacture their own food from carbon dioxide and water, using energy from sunlight that is trapped by the green pigment called **chlorophyll**. The process is called **photosynthesis**.

- Plant cells are surrounded by a rigid cell wall made of **cellulose**. Pressure within the cell caused by the entry of water keeps the cell firm and supplies rigidity to the plant.

- Plants have a complex reproductive cycle, involving various agents to bring about the processes of **pollination** and, later, fruit or seed dispersal.

- Most plants have only a few, but easily identifiable **organs** – leaves, flowers, stems and roots.

- Asexual reproduction, where a parent plant gives rise to may offspring without the involvement of gametes, is relatively common in plants.

- Although most plants in their familiar form have diploid nuclei, very few have the non-matching XY sex chromosomes.

NB Both plants and animals are made up of many cells (they are thus both described as being **multicellular**).

Classification within the animal kingdom

Animals either possess or do not possess a **vertebral column**.

> **TIP**
>
> Avoid the word 'backbone'. It is not a very accurate term as there are many bones in the vertebral column. Those possessing a vertebral column are called **vertebrates**. (Those without are called invertebrates.)

There are five groups of vertebrates:

- Fish
- Amphibians
- Reptiles
- Birds
- Mammals.

Fish

All fish share the following characteristics (Figure 1.1):

- A **skeleton** made of **bone** or of the more pliable material, **cartilage**

- A skin covered with **scales**

- **Fins** that present a large surface area to push against the water when swimming

- **Gills** for extracting oxygen from water and supplying it to the blood.

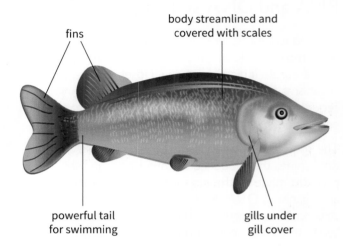

fins

body streamlined and covered with scales

powerful tail for swimming

gills under gill cover

Figure 1.1 A fish

Amphibians

Frogs, toads, newts and salamanders all have the following characteristics (Figure 1.2):

- A **soft skin** with no scales

- **Live** or can survive **on land** but always **return to water to lay eggs**

- Adults have **lungs** to breathe air

- Eggs hatch into larvae called **tadpoles** that live in water

- **Tadpoles** breathe using **gills**

- Tadpoles change into adults by metamorphosis (metamorphosis = a change in form and feeding habits from larva to adult).

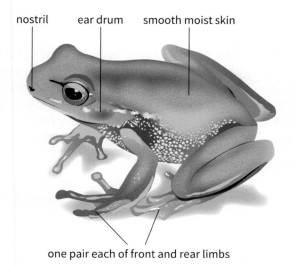

Figure 1.2 An amphibian (frog)

nostril   ear drum   smooth moist skin

one pair each of front and rear limbs

Reptiles

Lizards, snakes, tortoises and turtles all have the following characteristics (Figure 1.3):

- A tough, **dry**, **scaly skin**
- Lay **eggs** with leathery **shells on land**
- Have **lungs** for breathing air.

dry scaly skin

Figure 1.3 A reptile (crocodile)

Birds

Birds including the flightless birds (such as the ostrich) have the following characteristics (Figure 1.4):

- Skin covered with **feathers**
- Forelimbs modified to form **wings**

- A **beak** for feeding
- Have **scales** on their legs and toes
- **Lungs** for breathing
- Lay **hard-shelled eggs** on land
- Maintain body at a **constant temperature** – usually **above atmospheric temperature**.

> **TIP** Use the terms 'warm-blooded' and 'cold-blooded' with care. They are not very helpful terms, since some reptiles when basking in the sun may have a blood temperature higher than that of birds.

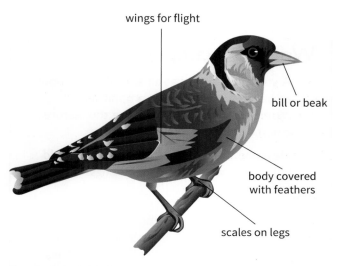

wings for flight

bill or beak

body covered with feathers

scales on legs

Figure 1.4 A bird

Mammals

Mammals, including kangaroos, cows, whales and human beings, have the following characteristics (Figure 1.5):

- Have **hair** on at least some part of the skin
- Internal fertilisation and **internal development** of the embryo
- Young ones fed on **milk from mammary glands**
- **Lungs** for breathing
- Maintain body at a **constant temperature** ('warm-blooded' – but sometimes not as warm as the surrounding atmosphere).

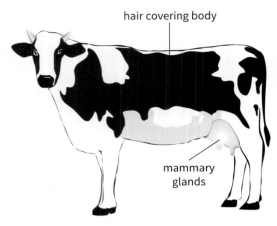

hair covering body

mammary glands

Figure 1.5 A cow

Test yourself by writing at least two characteristic features of each of the five vertebrate classes.

Worked example

a A student is told that an animal he is about to be shown is either an amphibian, a reptile or a mammal. Describe the features common these three groups.

b Describe the external features that would indicate to which group it belongs.

Answer

a Since these are all vertebrates, then the question is asking for vertebrate features shared by the three groups. All vertebrates have a vertebral column (avoid calling it a 'backbone') and a bony skeleton. They will possess eyes, a mouth, two front legs ('arms') and two rear legs. They will all have the following internal organs: lungs, heart, blood vessels, liver, kidneys and an alimentary canal.

b If it is an amphibian, it will have a soft, smooth and most probably moist skin. If it is a reptile, its skin will be tough, dry and will be covered with scales. If it is a mammal, it will have a skin covered, or partly-covered, with hair. It is likely to be warm to the touch. If it is a female, then it will have at least one pair of mammary glands on the front of its thorax (chest).

(Note that part (b) asks about external features. Mention of internal or external fertilization, laying of unprotected eggs in water that hatch into tadpoles (amphibian) or of shelled eggs on land (reptiles) are facts that are not relevant to the question.)

The invertebrates

These are the animals that **do not have vertebral columns**. Like the vertebrates, they are divided into **phyla** (the plural of 'phylum'), but such is the diversity of the invertebrates that they include over **30 different phyla**. The largest group (phylum) of invertebrates, by far, is the **arthropods**.

The arthropods

These include several Classes of which the largest and better-known ones are:

- The insects
- The crustaceans (crabs and lobsters)
- The arachnids (spiders)
- The myriapods (centipedes and millipedes).

All arthropods have the following features:

- They have **segmented bodies**.
- They have limbs with clearly visible joints.
- They have an **exoskeleton** (i.e. a skeleton on the outside of the body). (Muscles are attached internally to the exoskeleton – the opposite of ourselves, where muscles are attached externally to our endoskeleton.)
- The exoskeleton is composed of the chemical **chitin**. (See fungi.)

The insects

In addition to the characteristics of arthropods listed, insects have the following features (Figure 1.6):

- The **body** is divided into **three parts** – head, thorax and abdomen. The head, thorax and abdomen are not segments.
- They have **three pairs** of (jointed) **legs** – attached to the thorax.
- They usually have **wings** – one or two pairs attached to the thorax.
- They have **one pair** of **antennae** – attached to the head.
- They have **compound eyes** – each one with hundreds of small units called ocelli.
- **Breathing** is through **small holes** (spiracles), occurring in pairs, one each side of the abdominal segments and two on the thoracic segments. The spiracles lead into branched tubes called tracheae.

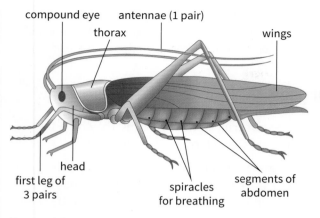

compound eye antennae (1 pair)

thorax   wings

first leg of 3 pairs

head

spiracles for breathing

segments of abdomen

Figure 1.6 An insect

The crustaceans

These live mostly in water (e.g. a lobster) and those that live on land (e.g. some crabs) live in damp places.

As well as the characteristic of arthropods, crustaceans have other features: (Figure 1.7):

- There are **two pairs** of **antennae** that are attached to the head.

- There are **three pairs** of **mouthparts** that, with the antennae, make up the **five** pairs of appendages attached to the head.

- The **exoskeleton** is often strengthened with **calcium salts**. (This protects the animal from predators, but can make the animal very heavy. The additional mass is supported by the water in which most crustaceans live.)

- The head and thorax are often joined to form the cephalothorax.

- The abdomen often has a **pair of limbs** on **each segment**, which are modified for many purposes, but often for swimming.

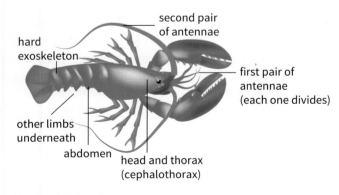

second pair of antennae

hard exoskeleton

first pair of antennae (each one divides)

other limbs underneath

abdomen

head and thorax (cephalothorax)

Figure 1.7 A lobster

Chapter 1 Classification

The arachnids

This group includes the spiders and the scorpions and, as well as those features possessed by arthropods, the arachnids also have the following (Figure 1.8):

- A **body** divided into two parts (the head and thorax, called the cephalothorax, and the abdomen).

- **Four pairs** of jointed legs joined to the cephalothorax.

- **No antennae.**

Figure 1.8 A spider

The myriapods

Myriapod means 'countless legs' and includes the centipedes and millipedes. As well as possessing the features common to arthropods, they also possess (Figure 1.9):

- **One pair** of **antennae.**

- **One** or **two pairs** of **legs** attached to **every segment**.

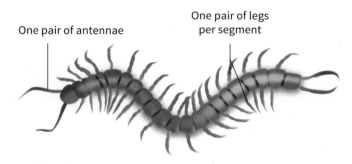

One pair of antennae

One pair of legs per segment

Figure 1.9 A centipede

7

Progress check 1.2

1 Name the chemical found in the exoskeleton of arthropods.

2 Which animals possess visible body segments?

    A   birds

    B   insects

    C   mammals

    D   reptiles

Other features present in cells

As well as cytoplasm, DNA and a cell membrane, the cells of all truly living organisms also contain within their cytoplasm structures ('organelles') called **ribosomes**.

Ribosomes are:

- about 20 nm in diameter (1 nm = 1 millionth of a millimetre)

- the place where amino acids are joined together to make proteins.

- made of protein and the nucleic acid RNA.

Some of the proteins made will be **enzymes** and amongst the enzymes will be those used in the process of **anaerobic respiration**. This occurs in the cytoplasm of all cells with the release of a relatively small amount of energy. If oxygen is available within the cell, then the end-products of anaerobic respiration will be further broken down (oxidised during **aerobic respiration**) to release greater amounts of energy.

The main features used in classifying viruses, prokaryotes (bacteria), protoctists and fungi

Since a microscope is required to study viruses and prokaryotes, protoctists and also for some **fungi** – such as yeast – they are referred to as **microorganisms**.

Occupying a position below the plants and animals in the evolutionary tree, the prokaryotes, protoctists and the fungi each form their own kingdoms (see earlier).

Viruses

Viruses are **not truly** living organisms. They have the following main characteristics (Figure 1.10):

1 They are **less than 300 nm** in size – around 50 times smaller than a bacterium. (1 nm, or nanometre, is 1 thousand millionth of a metre). They can be seen only with an **electron microscope**.

2 They contain **nucleic acid** (**DNA** or **RNA**).

3 The nucleic acid is surrounded by a **protein coat** (known as the **capsid**).

4 They can **reproduce** only inside living (**host**) cells.

5 Since they are **parasites**, they cause disease (i.e. they are pathogenic). Examples of diseases caused by viruses are influenza, measles and AIDS.

6 Viruses are **not** affected by **antibiotics**.

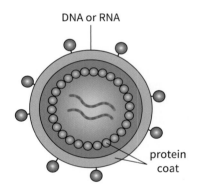

Figure 1.10 A virus

Their extremely small size allows them to be easily transmitted from host to host in very considerable numbers, both by air currents as well as by contact. The protein coat gives the nucleic acid considerable protection. Once inside a living host cell, they take over the host cell's metabolism and use it for their own reproduction. Some viruses (e.g. influenza virus) have a high **mutation** rate; thus a person may recover from flu, but still fall victim to the next epidemic caused by a mutated strain of the virus to which they have no immunity.

Prokaryotes (or bacteria)

Prokaryotes are truly living organisms, with the following characteristics (Figure 1.11):

1　They have a size in the range of **0.5–5 μm** (1 μm = 1/1000 mm).

2　They are unicellular (made of one cell only).

3　They have **no true nucleus** (their DNA lies 'loose' in the cytoplasm).

NB Some prokaryotes contain a loop of DNA called a plasmid – a feature sometimes employed in biotechnology.

4　They have a cell wall.

5　They may be (pathogenic) parasites or they may be saprotrophs. Some may be involved in nitrogen fixation and **denitrification**. (See the Nitrogen Cycle.)

Pathogenic = disease-causing; Saprotrophic = feeding on dead organic matter causing it to decay.

6　They are **killed** by **antibiotics**.

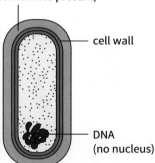

Figure 1.11 A bacterium

They can form resistant spores that are easily carried by air currents and by contact. Once within a suitable **substrate**, they reproduce quickly – dividing every half an hour.

Progress check 1.3

1　In suitable conditions, one bacterium could become how many in 5 hours?

2　What is the purpose of ribosomes?

Protoctista

These are a group of largely microscopic, truly living organisms.

1　They are mostly unicellular but some are multicellular.

2　They are either free-living or parasitic.

3　They have **aerobic** respiration involving mitochondria.

4　Unlike the prokaryotes, structures that lie in their cytoplasm are surrounded by **membranes**.

5　They have true **nuclei** (i.e. they are eukaryotes).

6　They reproduce both sexually and asexually.

7　They are grouped into **three categories**: animal-like (protozoa) that have animal-like nutrition; plant-like (algae) that feed by photosynthesis and fungus-like.

Fungi

Fungi are usually much larger organisms, mostly visible to the naked eye. For example: yeasts, moulds and mushrooms. They have the following characteristics (Figure 1.12):

- They have **no chlorophyll**. (They release enzymes to digest large molecules externally, then absorb the soluble products.) They are thus **parasites** or saprotrophs.

- They have a 'cell' wall made of **chitin**.

- They are usually made of a large number of tubular threads (**hyphae**) intertwined to form a **mycelium**.

- Hyphae are not divided into individual cells. The lining of **cytoplasm** has **many nuclei** and the central space in the hyphae is a vacuole full of (vacuolar) sap.

- If they store carbohydrate, they store glycogen.

- They reproduce by producing **spores**.

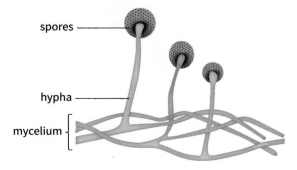

Figure 1.12 A fungus

The branching mycelium of a fungus ensures that the maximum amount of food substance (substrate) is digested as quickly as possible before it dries up, or is digested by bacteria. Fungi are important in the decay of dead, organic matter. Fungal spores are light and easily carried by air currents from one substrate or host to another.

1.04 Classification of the plant kingdom

In plant classification, the term 'division' is often used instead of phylum, but like phyla, divisions are divided into classes. Two major divisions of plants are the **ferns** and the **flowering plants**.

Characteristics of ferns

Ferns have the following characteristics (Figure 1.13):

- They are green photosynthesising plants.
- They have conducting tissue (**xylem** and **phloem**) forming veins.
- They have, often compressed, stems called **rhizomes**.
- They do **not** produce flowers.
- Instead, they produce **spores** that are light and easily carried away by the wind.
- Spores are released from **spore cases** (sporangia) that are found on the lower surfaces of fronds.
- **Frond** is the term for the leaves of ferns.

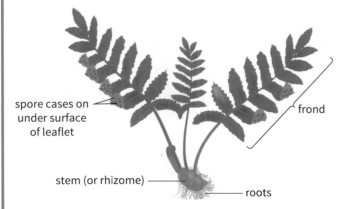

spore cases on under surface of leaflet

frond

stem (or rhizome)

roots

Figure 1.13 A fern plant

By far the largest division of the plant kingdom is the **flowering plants** that are sub-divided into two classes:

the **monocotyledons** and the **dicotyledons** (usually shortened to 'monocots' and 'dicots').

(NB The dicots have more recently been divided into several separate smaller groups, the most important of which is called the 'eudicotyledons'.)

1.05 Classes of flowering plant

Monocotyledons

Mono = one, cotyledon = a leaf that forms part of the structure of the seed.

This class includes the grasses, cereals, lilies and orchids, all of which share the following characteristics (Figure 1.14):

- **One** cotyledon inside each seed
- Leaves that are **narrow** and **strap-like**
- Leaves that have **parallel veins**
- A mass of equally sized (**fibrous**) roots
- **Flower** parts that are usually arranged in **threes** (i.e. three petals etc.).

Figure 1.14 A monocotyledon (*Iris*) with thin strap-like leaves that have parallel veins

The (eu)dicotyledons

This class includes cabbage, hibiscus, geranium and sweet potato, all of which have features that differ from those of the monocotyledons mentioned earlier (Figure 1.15).

- **Two cotyledons** are present inside each seed. Not only do these become the first photosynthesizing leaves when the seedling emerges above ground, but generally store food used during the process of seed germination.

- The leaves are **broad**.

- The leaves have **branched veins** usually radiating from a central thicker vein called the midrib with the branches linked by a network of veins.

- **Fewer,** thicker **roots** which are often joined to one long central root called the **tap root**.

- **Flowers** have parts usually arranged in **fours** or **fives**.

Figure 1.15 Flower of a dicotyledonous plant (flax)

Progress check 1.4

1 Which chemical is found in the cell walls of fungi?

   A   cellulose

   B   chitin

   C   glycogen

   D   protein

   →

2 An organism possesses xylem but never produces flowers. Which of the following will it be?

   A   a dicotyledon

   B   a fern

   C   a fungus

   D   a monocotyledon

3 An organism has three petals. Which of the following will it be?

   A   a dicotyledon

   B   a fern

   C   a fungus

   D   a monocotyledon

1.06 Dichotomous keys

Dichotomous means cutting (or dividing) into two.

Organisms are often identified using a book of illustrations. This is possible only if such a book is available, and this is the case only with certain organisms such as common plants, birds and butterflies. Even when such a book is available, identification will rely on the accuracy of the illustration, and it can be a time-consuming process if the organism is at the back of the book! For these reasons, biologists use **dichotomous keys**.

A dichotomous key consists of a series of questions. Each question has two alternative answers. Depending on which answer is chosen, the user is directed to the next question. Thus, by starting at the first question, and then by a process of elimination, a specimen may be identified.

This process is reliable because it directs the user to observe particular characteristic features. Also it is quicker since, at each question, possible alternatives are eliminated.

Dichotomous keys are usually presented in the following format. The example chosen is a key in its simplest form – namely to identify the kingdom into which an organism should be placed. More detailed keys are used when determining precisely to which **species** from many within the same genus an organism belongs.

1	Is it unicellular (i.e. made of only one cell)?	Yes	go to 2
		No	go to 3
2	Does it have a nucleus?	Yes	protoctist
		No	bacterium
3	Does it have hyphae?	Yes	fungus
		No	go to 4
4	Does it have cell walls?	Yes	plant
		No	animal

When identifying one organism from amongst a large number of possibilities, the most effective dichotomous key asks questions that each time divides the remaining possibilities into roughly equal halves. In this way, half the possible organisms are discarded at each step.

Progress check 1.5

1 Six different geometrical shapes, identified by the letters A to F are shown in Figure 1.16.

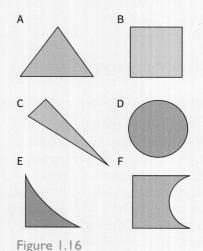

A B

C D

E F

Figure 1.16

Construct a dichotomous key to identify the six shapes.

2 Now construct a key to identify the shapes using different features from those you have used already.

You should now find that several different keys may be constructed, all of which may be perfectly suitable for the purposes of identification.

Chapter summary

- [ ] You now know the characteristics of living organisms.

- [ ] You have learnt how to use the binomial system of naming organisms.

- [ ] You are able to list the five classes of vertebrate and know how to distinguish between them.

- [ ] You have learnt the differences between the microorganisms viruses and bacteria and how they differ from fungi.

- [ ] You are now able to write down the names and characteristics of the four phyla of invertebrates and the four classes of arthropods that are described.

- [ ] You have learnt the differences between the two classes of flowering plant.

- [ ] You have learnt how to use and also to construct a dichotomous key for identifying organisms.

Exam-style questions

1 Figure 1.17 shows six arthropods.

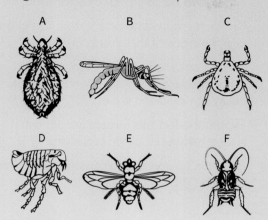

A   B   C

D   E   F

Figure 1.17

Use the key to identify each of the arthropods. Copy Table 1.1 and write the name of each arthropod in the correct box. As you work through the key, put a tick (✔) in the boxes to show how you identified each arthropod.

As an example, the appropriate boxes for arthropod A have been ticked for you.

Key

			Arthropod
1	a	segments on abdomen clearly visible	go to 3
	b	segments on abdomen not clearly visible	go to 2
2	a	3 pairs of legs	*Pediculus*
	b	4 pairs of legs	*Ornithodorus*
3	a	wings present	go to 4
	b	wings absent	*Pulex*
4	a	wings clearly longer than abdomen	*Musca*
	b	wings not clearly longer than abdomen	go to 5
5	a	antennae curved	*Periplaneta*
	b	antennae straight	*Anopheles*

	1a	1b	2a	2b	3a	3b	4a	4b	5a	5b	name of arthropod
A		✔	✔								*Pediculus*
B											
C											
D											
E											
F											

[10]

Table 1.1

2   Figure 1.18 shows a centipede.

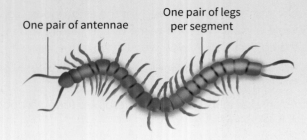

One pair of antennae

One pair of legs
per segment

Figure 1.18

The centipede is a myriapod, one of the four groups
of arthropod. It is a carnivore that lives on land and,
compared with most other arthropods, its outer
body covering is thin and permeable.

a   Name **two** other groups of arthropod.

For each group, state one feature found only
in arthropods of that group. [4]

b   Suggest and explain two reasons why
centipedes are often found under stones,
decaying wood and leaves. [4]

3   a   Explain the fact that viruses are almost
always harmful to other organisms. [6]

b   Describe how viruses differ structurally from
bacteria. [5]

Cells

2.01 Cell structure and organisation

The basic unit of life is the **cell**. The simplest living organisms have one cell only. Such organisms are described as **unicellular**. Bacteria are examples of unicellular organisms. Most other living organisms have many cells. They are described as **multicellular**.

All cells have the following structural features in common (Figure 2.1 and Figure 2.2):

1   **Cell surface membrane:** This surrounds the cell and **controls** the passage of substances into and out of the cell. One of the most important of those substances is water. All other substances that pass through, do so in **solution**.

2   **Cytoplasm:** A jelly-like substance in which the chemical reactions of the cell (**metabolic reactions**) take place and which contains the nucleus.

3   **Nucleus:** This contains a number of **chromosomes** made of the chemical DNA.

NB Cytoplasm and nucleus together may be referred to as **protoplasm**.

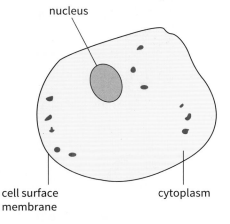

nucleus

cell surface membrane

cytoplasm

Figure 2.2 Animal cell (liver)

Plant cells have the following additional structures (Figure 2.3):

1   A large, central **vacuole**: a space full of **cell sap** (thus sometimes called the **sap vacuole**), which is a solution mostly of sugars. It is separated from the cytoplasm by the **vacuolar membrane**. (Plant cells undergoing cell division do **not** have a vacuole.)

2   **Cell wall**, which is a 'box' made of **cellulose** that contains the cell.

3   **Chloroplasts**. These are present only if the cell is involved in the process of **photosynthesis**. These are small bodies lying in the cytoplasm, which are green in colour because of the pigment **chlorophyll** that they contain.

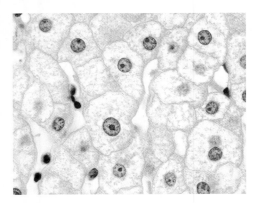

Figure 2.1 Stained liver cells

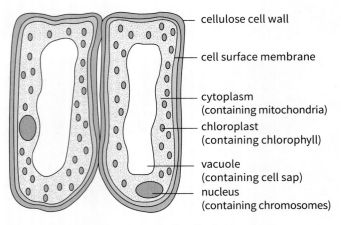

cellulose cell wall

cell surface membrane

cytoplasm
(containing mitochondria)

chloroplast
(containing chlorophyll)

vacuole
(containing cell sap)

nucleus
(containing chromosomes)

Figure 2.3 Palisade mesophyll cells from a leaf

In plants, the cell membrane is not usually easily visible as it fits tightly against the cell wall.

Progress check 2.1

1   Figure 2.4 shows a cell from the leaf of a plant.

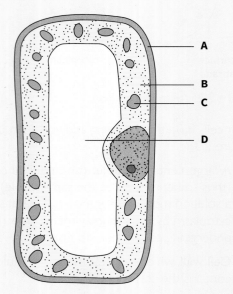

A

B

C

D

Figure 2.4 Palisade mesophyll cells from a leaf

Which labelled feature is also found in an animal cell?

2   Why do cells from the root of a plant rarely contain chloroplasts?

2.02 How the structural features in a cell are related to their functions

Cell surface membrane

This controls the passage of materials in and out of the cell since it is **partially** (or **selectively**) **permeable**. That means that it allows some chemicals to pass through it but not others. Some small molecules pass through minute pores in the membrane, and the process does not require energy. Other, usually larger, molecules pass through the membrane via specific pathways, and may pass through when there is a greater concentration inside compared with outside the cell. This process does require energy.

Cytoplasm

This is where the chemical reactions of the cell occur such as **respiration** and protein manufacture. It contains a range of small structures called **organelles** for example, the **chloroplasts** in plant cell cytoplasm where photosynthesis and carbohydrate (often starch in plants) storage takes place.

Further examples of structural features inside cells are:

- Mitochondria, the largest of which may just be visible using a light microscope and which are involved in the storage of energy (in the form of ATP) released by respiration (Figure 2.5, 2.6). Although mitochondria are present in almost all cells, they are **not present** in the cells of **prokaryotes**. A mitochrondrion has many folds (called cristae) on its inner walls, and it is on these folds that the process of **aerobic respiration** takes place. It is for this reason that mitochondria are referred to as the 'power houses' of the cells.

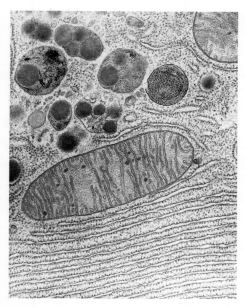

Figure 2.5 Photomicrograph of a mitochondrion with endoplasmic reticulum beneath with ribosomes on its walls

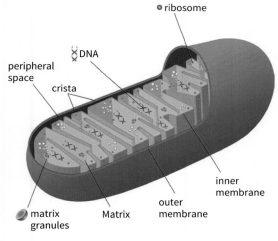

Mitochondrion

Figure 2.6 Diagram of a mitochondrion

- **Ribosomes** are where amino acids are linked together to form proteins. Ribosomes lie attached to the walls of a system of many-folded small tubes (microtubules) known as the **endoplasmic reticulum** – often shortened to **ER** (Figure 2.7). The ER runs throughout the cytoplasm of cells. Endoplasmic reticulum with ribosomes attached is known as rough endoplasmic reticulum.

- **Vesicles** are spherical bodies that break away from the endoplasmic reticulum and which contain proteins that have been made by the cell. They are used to transport and, sometimes, export these proteins.

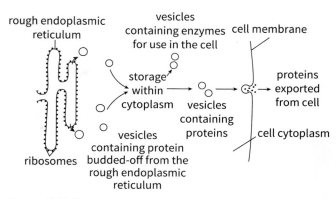

Figure 2.7 Endoplasmic reticulum and vesicles

Mitochondria are present in the largest numbers in cells that release large amounts of energy. For example, those involved in active uptake (absorption) of chemicals and in movement, for example, **sperm** cells. Ribosomes are made of RNA and protein.

Nucleus

This can be regarded as the organelle that controls the cell's activities. It contains thread-like structures called **chromosomes** that are made largely of DNA.

DNA forms **genes** that are responsible for programming the cytoplasm to manufacture particular proteins.

When any cell other than those that produce reproductive cells (**gametes**) divides, its nucleus does so by a process called **mitosis** during which each chromosome forms an **exact replica** of itself. The two cells formed are therefore genetically identical both with themselves and with the original cell.

Cell wall

Plant cells are enclosed in a 'box' of flexible but tough carbohydrate called **cellulose**, which is a **completely permeable** substance. Thus it does not in any way affect materials passing in and out, but it helps the cell to:

- keep its shape
- prevent the cell from bursting as it absorbs water
- maintain a pressure within the cell (**turgor** pressure) to keep the plant firm and upright.

Sap vacuole

The importance of this structure is to contain a solution more concentrated than the solution in the soil water around the plant. A more concentrated solution has a **lower water** potential and this causes water molecules to enter the plant from the soil by **osmosis**. (See Chapter 3, Osmosis.)

Similarities and differences between plant and animal cells are shown in Table 2.1.

TIP

Remember that $1\,\mu m = \dfrac{1}{1000}\,mm$

	Animal cell	Plant cell
Similarities	cell membrane	
	cytoplasm	
	nucleus	
Differences	no sap vacuole	sap vacuole
	no cell wall	cell wall
	no chloroplasts	may have chloroplasts
	never stores starch	may store starch
	around 10–20 µm in diameter	around 40–100 µm in diameter

Table 2.1 Comparison of plant and animal cells

PRACTICAL

Make sure you are familiar with the procedure for preparing animal and plant cells for viewing under a microscope.

1   **To observe animal cells:**

Cut a cube of fresh liver, in section, approximately 1.5 cm square.

Scrape one of the cut surfaces of the cube with the end of a spatula (the end of a teaspoon would do).

Transfer the cells removed to a clean microscope slide. Add one drop of **methylene blue** (a suitable stain for **animal** cells) and one drop of glycerine.

Mix the cells, stain and glycerine together gently and leave for 30 seconds. (This time can be adjusted according to the depth of staining required.)

Carefully place a clean, dry cover slip over the preparation, and then wrap a filter paper around the slide and cover slip.

Place the slide on a bench and press hard with your thumb on the filter paper over the cover slip. The filter paper should absorb any surplus stain and glycerine, and the slide is then ready for viewing with a microscope (medium to high power).

Figures 2.1 and 2.2 show structures that should be visible.

2   **To observe plant cells:**

Peel off the dry outer leaves of an onion bulb.

Remove one of the fleshy leaves beneath.

Preferably using forceps, but fingers would do, peel away the outer skin-like covering (**epidermis**) of the fleshy leaf.

Place three drops of dilute iodine solution on a clean, dry microscope slide (iodine solution is a suitable, temporary stain for **plant** cells).

Transfer a small piece of the epidermis (a 50–75 mm square is large enough) to the iodine solution. (Make sure it lies flat and is completely covered by the iodine solution.)

Carefully place a glass cover slip on top of the preparation, remove any excess liquid with a piece of filter paper and transfer the slide to the stage of a microscope.

The structural features shown in Figure 2.8 should be visible (owing to the large size of the onion cells, it may not be necessary to use the high power of your microscope).

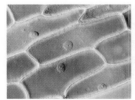

cellulose cell wall
cell surface membrane
nucleus
cytoplasm

Figure 2.8 Photomicrograph and labelled drawing of onion cells

2.03 Levels of organisation

Specialised cells, tissues and organs

In unicellular organisms, one cell must be able to carry out all the functions of a living organism. In multicellular organisms, cells are usually modified to carry out one main function. The appearance of the cell will vary depending on what that main function is.

Thus, there is a relationship between the structure and the particular function of a cell.

Examples of this relationship are discussed here.

Ciliated cells

Function

To sweep mucus, in which dust and bacteria are trapped, up the bronchi and trachea towards the throat where it is swallowed.

How they are adapted to this function

Ciliated cells are found lining the walls of the trachea (wind-pipe) in the respiratory tract. Each cell bears a fringe of minute projections (**cilia**). The singular of cilia is cilium.

The cilia perform an upward-beating motion that carries the mucus, made and released by neighbouring cells, upwards like a 'moving carpet'.

TIP

Remember that cilia do not form a network to 'trap' dust and bacteria.

Root hair cells

The tip of a root with its many **root hair cells** is shown in Figure 2.8.

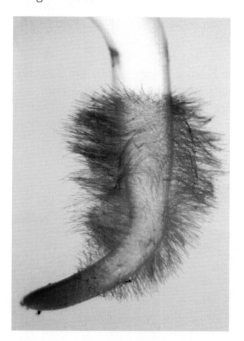

Figure 2.9 A root tip showing root hairs

Function

The absorption of water and mineral **ions** (salts) from the soil.

How they are adapted to this function

The outer part of the cell wall of each **root hair cell** (i.e. the part in direct contact with the soil) is in the form of a long, tubular extension (the root hair).

This root hair:

- is able to form a very close contact with the water film surrounding many soil particles and

- it **greatly increases the surface area** of the cell available for uptake of water and ions.

Xylem vessels

Functions

1 To **conduct** water and ions (**dissolved** salts) from the roots to the stem, leaves, flowers and fruits.

2 To provide **support** for the aerial parts of the plant.

How they are adapted to these functions

Conduction

Xylem vessels are elongated dead cells forming long, narrow tubes, stretching from the roots, via the stem, to the leaves. They are stacked end to end like drain pipes.

Support

1   Their walls have been strengthened by the addition of the chemical lignin. (As the lignin in the walls builds up, it eventually kills the xylem vessels. There is then no layer of cytoplasm to restrict the flow of water and dissolved salts.)

2   Xylem vessels are part of the vascular bundles (Figure 2.9), which run the entire length of the stems of plants thus resisting bending strains caused by the wind.

TIP

Vascular bundles help to strengthen a stem since they work like iron reinforcements in concrete pillars.

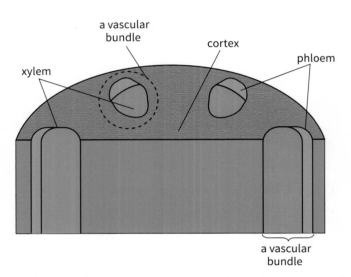

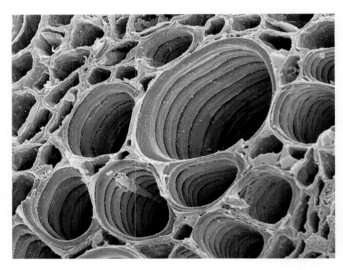

Figure 2.10 3D section of stem showing xylem as part of the vascular bundles and a photomicrograph of xylem vessels

Palisade mesophyll cells

The inner structure of a plant leaf includes two specialised types of cell known as mesophyll cells.

The lower cells are called spongy cells and the upper layer are the palisade mesophyll cells (Figure 2.3). The palisade mesophyll cells form the main **photosynthesising tissue** of the leaf, helped by the fact that they:

*   contain the greatest number of chloroplasts

*   they are the first to receive the Sun's rays as they enter the leaf

*   their chloroplasts are able to move, within the cytoplasm of the cell, towards the upper surface of the leaf to receive more sunlight.

Nerve cells (or neurones)

These may be long, thin cells that carry electrical impulses from the receptor organs to the central nervous system, or from the central nervous system to the organs that are required to respond (the effectors) (see Chapter 13, Figure 13.3). Neurones in the central nervous system, especially the brain, are more compact.

Red blood cells

Function

To carry **oxygen** around the body.

How they are adapted to this function

1 The cytoplasm of **red blood cells** contains the pigment **hemoglobin**, which combines (in the lungs) with oxygen to become **oxyhemoglobin**.

2 They are small (**7 µm × 2 µm**) (Figure 2.11), and there are many of them. This gives them a **very large surface area** for oxygen absorption.

3 They have a **biconcave** shape, increasing their surface area for absorption still further.

4 They are flexible allowing them to be pushed more easily through **capillaries**.

Points 2, 3 and 4 are also relevant to their other function of carrying carbon dioxide.

surface view

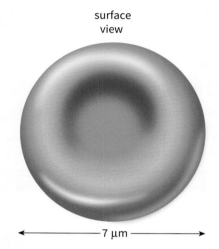

←———— 7 µm ————→

side view

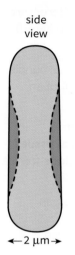

←2 µm→

sectional view

Figure 2.11 Red blood cells

Sperm and egg cells

Together known as **gametes**, these are the cells used in sexual reproduction. They are produced by a form of cell division (**meiosis**) that halves the number of chromosomes they possess, that is they become haploid. In this way, when they unite at **fertilisation**, the full number of chromosomes (the diploid number) is restored.

The sperm

This is the male gamete, it is made of:

- a head containing the haploid nucleus

- a middle piece that supplies energy for movement

- a tail that beats to provide propulsion as it swims through the female reproductive system to locate the egg cell.

Sperms are released in millions at a time to increase the chances of fertilisation. A human sperm is about 100 µm in length.

The egg cell (or ovum)

This is the female gamete with the following features:

- The haploid nucleus is contained within a large amount of nutritive cytoplasm.

- It does not possess the ability to move on its own, but is moved by the action of cilia and muscles in the female reproductive system.

- It is spherical and much larger than the sperm, measuring between 0.10 and 0.15 mm in diameter.

- Egg cells are released one at a time.

Progress check 2.3

1 In what way are cells in the trachea adapted to carry mucus up to the throat?

   A They are biconcave in shape.

   B They contain many fibrils.

   C They have strengthened cell walls.

   D They possess many cilia.

2 What features do root hair cells and red blood cells have in common and what common purpose do these features serve?

3 Which cells in a plant leaf have the greatest number of chloroplasts and why is this?

2.04 How cells combine to improve their efficiency

One cell working on its own would achieve very little in an individual plant or animal, thus it is usual to find many similar cells lying side by side and working together, performing the same function.

A tissue is therefore: **many cells with similar structure working together and performing the same (shared) function**.

Examples of tissues: xylem tissue in the vascular bundles of a plant, muscular tissue in the intestine wall of an animal.

Different types of tissue often work together in order to achieve a combined function. **Several tissues working together to perform specific functions** form a structure called an organ.

Examples of organs: the leaf of a plant – an organ for the manufacture of carbohydrates during photosynthesis, the eye of an animal – the organ of sight.

Several different organs may be necessary in order to carry out a particular function. **A group of organs with related functions working together in order to perform a particular body function** is called an organ system.

Examples of organ systems: the sepals + petals + stamens + carpels (i.e. the flowers) of a plant for reproduction; the heart + arteries + veins + capillaries in an animal, i.e. the circulatory system.

An **organism** is a collection of organ systems working together.

The increasing order of cell organisation found within any living organism is thus:

cell ➞ tissues ➞ organs ➞ organ systems ➞ organisms

Progress check 2.4

1   The following are structures found in living organisms:

P the eye

Q the muscles in the intestine wall

R a flower

Which shows the level of organisation in these three structures?

	Tissue	Organ	Organ system
A	P	Q	R
B	Q	R	P
C	R	P	Q
D	Q	P	R

2   Make a list of two different types, other than those mentioned in the text, of cell, tissue, organ and organ system. (Check with your teacher that you have correctly classified you chosen examples.)

Worked example

Explain how the terms cell, tissue and organ may be applied to the structure of a leaf.

Answer

It is easiest to begin with the fact that leaves are made up of a large number of cells. This would then be followed by some different examples of types of cells found in a leaf – for example, palisade cells and, if you have looked up leaf structure, you could include spongy cells, epidermal cells and cells found in the vascular bundles (e.g. in the xylem – which start out as living cells). The answer should then refer to the fact that the cells are modified for different functions – palisade cells contain chloroplasts for photosynthesis and xylem cells become modified for support and conduction.

The cells are found working beside many other similar cells, each one of which is modified in the same way and performs the same function. Such a group of cells forms a tissue – for example, the palisade tissue for making starch during photosynthesis (and the xylem tissue for bringing water and ions into the leaf and for supporting the leaf). A leaf thus contains a number of groups of different cells, each group performing its own job, but with each group performing one of the necessary functions of a leaf. Thus, the leaf is an organ, since it have several tissues working together performing specific functions.

The size of specimens

Biologists deal with specimens with a wide variety of sizes. If a specimen is rare, then a description of it is invaluable to other biologists in particular (and other people in general). A drawing is often the best way to record observations, but such a drawing is of little value if it does not give an indication of the size of the object being observed.

All drawings of biological specimens should, therefore, include a reference to the magnification of the drawing.

> **TIP**
>
> When drawing, measure in mm – avoid cm – so give a measurement as 5 mm rather than 0.5 cm, 50 mm rather than 5 cm.
>
> Clearly state the **linear** dimension measured from the drawing over the matching (linear) dimension measured on the specimen when calculating the magnification, e.g. 42 mm/18 mm = × 2.3.

Remember, if the specimen is large, then the magnification will almost certainly be less than 1.

- Magnification should be written, e.g. × 4.4.

- Don't round off too much, × 4.4 is not × 4.

- No calculated magnification should include more significant figures than the least accurate of the measurements used to calculate it, e.g. 61 mm/14 mm = × 4.4 (not × 4.3571429).

- If the subject is a **photograph** it may also have a **stated magnification** that needs to be taken into account – and shown – in the calculation. Example: a photograph may bear the caption '× 4' thus the measurement taken from the photograph must be divided by four in your calculation, e.g. 14 mm/(12 mm/4) = × 4.7).

(Remember that diagrams and photographs of specimens as seen using a microscope may have dimensions in micrometres (μm) and 1 micrometer = 1000th of a millimetre.)

Chapter summary

- [ ] You have learnt the component parts of a cell.

- [ ] You have also learnt how plant and animal cells are similar and how they differ.

- [ ] You know how cells, tissues, organs and organ systems are related.

- [ ] You have learnt how cells are modified to perform different functions.

- [ ] You have also learnt what is required when drawing a specimen.

Exam-style questions

1 Figure 2.12 shows two cells.

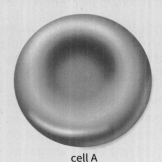

cell A   cell B

Figure 2.12

a i) State where, in a human, Cell A would normally be found. [1]

ii) Name the chemicals responsible in each case for the colour of the two cells: Cell A and Cell B. [2]

b Using only words from the list provided, complete the statements about Cell B.

air spaces	cellulose
chloroplasts	membrane
mitochondrion	nucleus
starch	vacuole
cell wall	cytoplasm

Photosynthesis occurs in the _____ situated in the _____ of the cell. It uses carbon dioxide absorbed from the _____ in the leaf. The _____, which is made of _____, under pressure from the _____ in the _____, helps the cell to maintain its shape. [7]

[Total 10]

2 Figure 2.13 shows the structures that produce urine and excrete it from the body.

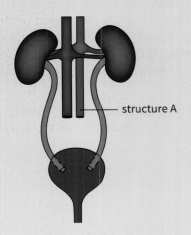

structure A

Figure 2.13

a In Table 2.2, put a tick (✓) in the box next to the term that states the level of organisation of structure A.

organism	
organ system	
organ	
tissue	
cell	

[1]

Table 2.2

b With reference to the kidney, ureters and bladder, explain the difference between the terms *organ* and *organ system*. [6]

3 Explain how a plant's xylem is particularly suited to its functions. [10]

Movement in and out of cells

3.01 Diffusion

Chemicals must be able to move from one part of a cell to another, into and out of a cell and from one cell to another if an organism is to remain alive.

It is an advantage if this movement requires no effort (or, more correctly, no 'expenditure of energy') on the part of the organism, and, so long as there is no obstruction, chemical molecules carry out this process by diffusion.

Diffusion relies on the fact that all molecules are in a constant state of random (kinetic) motion, but before diffusion can occur, there must be a concentration gradient of the molecules, that is, a region of their (relatively) high concentration immediately beside a region of their (relatively) low concentration.

Diffusion can then be defined as **the net movement of molecules and ions from a region of their higher concentration to a region of their lower concentration, down a concentration gradient as a result of their random movement.**

The energy that causes the movement of chemicals is the kinetic energy that results from the movement of molecules and ions.

3.02 Examples of diffusion

Living organisms make regular use of the diffusion of gases or chemicals in solution (solutes). Only those molecules small enough to pass through the cell surface membrane are able to pass in and out of a cell.

Some examples of diffusion are described here.

In plants

1 The movement of **carbon dioxide**, first as a gas from the atmosphere into a leaf, then in solution, from the water film surrounding the mesophyll cells in a leaf, to the **chloroplasts** during photosynthesis.

2 The movement of **water vapour** and **oxygen** (both in **gaseous** form) released from the water film surrounding the mesophyll cells inside a leaf, and both then passing through the intercellular spaces of the leaf, and out through the stomata. Oxygen first diffuses in solution from the photosynthesising cells into the water film surrounding the cells.

The loss of water vapour in this way is called transpiration.

In animals

1 The movement of **oxygen** after it has dissolved in the moisture lining the air sacs of the lungs through the walls of the air sacs (alveoli) into the blood.

2 The movement of **carbon dioxide**, in solution, from the cells through tissue fluid, into the blood in blood capillaries.

In all the examples given here – apart from water vapour, the substance that diffuses has first **dissolved in water** (i.e. it is the **solute** molecules that are diffusing).

Progress check 3.1

1 How many examples of diffusion can you think of – first in the world around you, then in plants and in animals? (Check your list with your teacher to see how many of your examples were correct.)

2 What causes molecules to move by diffusion?

The rate at which a substance diffuses is controlled be a number of different factors:

Chemicals are absorbed faster by diffusion:

- when the structure absorbing them has a large surface area compared with its volume. This is achieved by the absorbing structure being long and thin (e.g. in **root hair cells** of plants or the **villi** in the small intestines of mammals), or simply by being very small

- when the temperature is higher, increasing the amount of kinetic energy in the molecules involved

- when the concentration gradient across which the diffusion is occurring is greatest

- when less distance has to be travelled by the diffusing molecules. For this reason, respiratory surfaces, for example, the walls of the alveoli in the lungs that absorb oxygen and release carbon dioxide are only one cell thick.

PRACTICAL

Investigations of the factors that influence the rate of diffusion

I The effect of surface area

Apparatus: A petri dish
Three cork borers of different sizes
Two medium-sized beakers
Two dropping pipettes
Safety goggles and thin protective gloves

Materials: Agar tablets or powdered agar
Phenolphthalein pH indicator
1 mol/dm^3 hydrochloric acid
Bench dilute sodium hydroxide
A supply of water (preferably ionised or distilled) and a means of heating it.

Method:
Following the instructions on the bottle, dissolve some agar in hot water. When it has dissolved, using a pipette, add a few drops of phenolphthalein indicator to the agar solution. If the agar solution is alkaline, it will turn pink*; if not, using the second pipette, add drops of sodium hydroxide until it does so. Pour the pink agar solution into the petri dish and allow it to cool and set in a refrigerator.

*Colour-blind students may find this investigation easier if they use a pH indicator that does not include pink as a significant colour.

Wearing the gloves and goggles, fill the second beaker to a depth of at least 3 cm with the hydrochloric acid (the beaker should be large enough to very easily accommodate the agar cylinders mentioned later).

Using each cork borer, in turn, bore out three cylinders of agar from the petri dish.

(If you have trouble in removing the agar from the cork borers, you can, instead, using a sharp blade, cut the agar into three different-sized cubes, which you should measure carefully.)

Transfer the three cylinders/cubes to the beaker of hydrochloric acid and note the time.

Keep a constant watch on the cylinders/cubes and record the time at which each cylinder loses its pink colouration.

Calculate the volume and the surface area of each agar cylinder/cube.

The volume of each cylinder can be calculated by measuring the depth of the agar in the petri dish and the diameter of the cork borer and using the formula $\pi r^2 h$ (where h is the depth of the agar and r is half the diameter of the cork borer).

The surface area of each cylinder is calculated using the formula $2\pi r^2 + 2\pi rh$.

Results:
The larger the ratio of surface area to volume for the cylinders/cubes, the shorter time it takes for the pink colour to disappear.

Explanation: The greater the surface area to volume ratio, the more H$^+$ ions are able to diffuse from the hydrochloric acid into the agar blocks changing their pH thus the quicker the indicator loses its pink alkaline colouration.

→

2   The effect of temperature

The investigation of surface area previously can be adapted to demonstrate the effect of temperature on diffusion.

Apparatus: 1 cork borer
4 medium-sized beakers
(size dependent of the size of the cork borer)
A container of ice and water and a water bath set at 40 °C

Materials: As before

Method:
Prepare the agar as in Investigation 1 – The effect of surface area.

Wearing the gloves and goggles, fill three beakers with the hydrochloric acid each to a depth of 3 cm.

Place one of the three beakers in the container of ice and water, one on the bench at laboratory temperature and one in the water bath, leave for 10 minutes.

Bore three cylinders of agar each with the same cork borer (or, cut three identical cubes of agar, as described previously).

Transfer one of the cylinders/cubes to each of the three beakers and note the time.

Keep a constant watch on the cylinders/cubes and record the time at which each one loses its pink colouration.

Results and explanation:
The cylinder/cube in the water bath loses its colour first and the one in the ice and water is last to lose its colour. Thus diffusion occurs faster the higher the temperature.

3   The effect of concentration gradient

Investigation 2 can be adapted to demonstrate this effect.

Materials: As in 1 and 2 previously, but a supply of 2 mol/dm³ hydrochloric acid is required, which, by dilution, should be used also to provide a supply of 1 mol/dm³ and 0.2 mol/dm³ hydrochloric acid. (Although these concentrations are suggested, the investigation should work well enough with three other differing concentrations.)

Apparatus: As in Investigation 2, but no container of ice and water or a water bath are required.

Method:
Prepare the agar as in Investigation 1 – The effect of surface area.

Fill the three beakers to the same depth (around 3 cm), one with 2 mol/dm³, one with 1 mol/dm³ and one with 0.2 mol/dm³ hydrochloric acid. All three beakers should be kept at the same (laboratory) temperature.

Bore three cylinders each with the same borer, or cut three identical cubes of agar.

Transfer one of the cylinders/cubes to each of the beakers and note the time.

Keep a constant watch on the cylinders/cubes and record the time at which each one loses its pink colouration.

Results and explanation:
The agar in the most concentrated acid is first to lose its colour and last to lose its colour in the least concentrated. There is a steeper concentration gradient of H⁺ ions, the more concentrated the acid, hence there is also a faster rate of diffusion of the H⁺ ions.

4   The effect of distance

This can be easily investigated by using two glass containers of significantly different volume. Each is filled with water, then placed side by side and left for the water to become still. Two of three drops of a concentrated solution of potassium permanganate are added, at the same time, to both containers, and the time is noted. The containers are then observed until the water in each of them is uniformly purple in colour and the time for each is recorded.

Results:
It takes longer for the water in the larger container to become uniformly coloured.

Explanation: It is not just that the potassium permanganate has further to travel, but also that its concentration gradient is progressively declining as it spreads out in the container.

The effect of distance can also be demonstrated by spraying aerosol deodorant, or fly-killer at one side of a classroom, and asking students to note the time taken before they can smell it. The further away they are, then, perhaps not surprisingly, the longer it takes for the chemical to reach them (by diffusion through the air).

3.03 Osmosis

If we were to take a container, completely divided into two sections by a piece of cloth, and then, at the same time pour a dilute sugar solution into one side and a concentrated sugar solution into the other side, within several minutes, by diffusion, both the water molecules and the sugar molecules would move down their respective concentration gradients, until both sides were at the same concentration (Figure 3.1). The pores in the cloth would form no obstruction to the movement of the molecules in either direction.

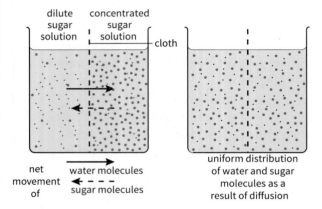

Figure 3.1 Sugar molecules moving one way, water molecules moving in the opposite direction until both sides were at the same concentration

If we now carry out a similar experiment but this time, instead of the piece of cloth, we separate the two sides of the container with a membrane with microscopic holes in it (Figure 3.2) – so small that they allow the passage of water molecules but not the sugar molecules, then water molecules will diffuse down their concentration gradient while the sugar molecules stay where they are. This **specialised** case of diffusion is called **osmosis** and the separating membrane is described as partially permeable.

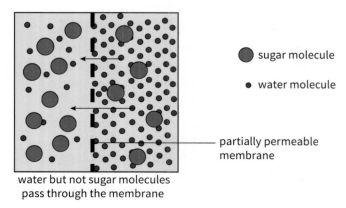

Figure 3.2 Partially permeable membrane allowing water but not sugar molecules to pass through

All cell membranes are partially permeable, therefore water will move into or out of cells by osmosis depending on the concentration gradient of water molecules inside and outside the cell.

Dilute solutions, which have a relatively large number of water molecules, are said to have a **high water potential** (i.e. a high concentration of water molecules). **Concentrated** solutions, with fewer water molecules, are said to have a **low water potential**.

During osmosis, since water molecules are in a constant state of motion, they will be moving through the partially permeable membrane in both directions, but always more will be moving from where they are in greater concentration towards the lower concentration.

Osmosis can then be defined as **the net movement of water molecules from a region of higher water potential (a dilute solution) to a region of lower water potential (a concentrated solution), through a partially permeable membrane**.

The cell sap of root hair cells has relatively **low** water potential. Soil water has a relatively **high** water potential. Thus water will move into the vacuole of root hair cells by **osmosis** (Figure 3.3). The cell wall offers **no obstruction** to the passage of water, since **cellulose**, of which it is made, is **completely permeable**. But, where the walls of neighbouring cells touch, water can pass into the root by simple diffusion – through the cellulose of the cell walls.

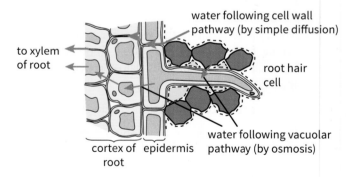

Figure 3.3 Uptake of water by osmosis (vacuolar pathway) and by diffusion (cell wall pathway)

Once water molecules have entered the root hair cell, their presence **increases** the water potential of that cell, compared with the next cell in towards the centre of the root. **Osmosis** will thus cause water to move into that cell from the root hair cell. This process continues until water molecules reach the **xylem** vessels in the centre of the root and are then transported away to the stem. This can be demonstrated in the laboratory using a piece of Visking tube that has been tied tightly at its lower end, that contains sucrose solution, and is tied tightly at its upper end to a piece of glass tubing. The apparatus is then supported in a beaker of water as shown in Figure 3.4.

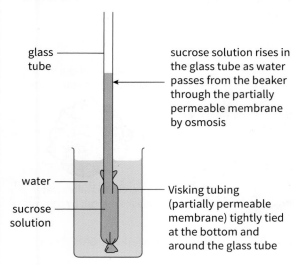

Figure 3.4 A demonstration of osmosis using an artificial partially permeable membrane

---

Progress check 3.2

1   What part of a cell is partially permeable?

2   What is the difference between the diffusion and osmosis?

The effect of osmosis on plant cells

The intake of water by osmosis

Water will always tend to enter plant (root) cells by osmosis, since soil water will always be likely to have a higher water potential than cell sap (Figure 3.5). As the vacuole thus increases in volume, it increasingly presses the cytoplasmic lining of the cell against the flexible, box-like cell wall.

(This pressure is called **turgor** pressure and it gives plant cells a firmness called **turgidity**.)

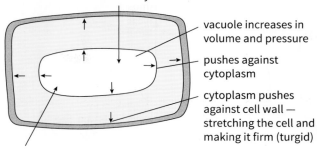

Figure 3.5 Plant cell in water

The increase in pressure resulting from osmosis can be demonstrated as shown in Figure 3.5 using a tightly-tied bag made of Visking tube, filled with sugar solution and placed in water for 20 minutes (Figure 3.6).

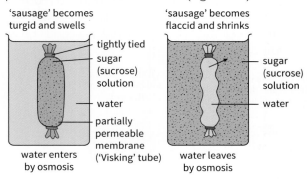

Figure 3.6 A firm, turgid, Visking sausage and one that has become flaccid

The pressure caused by the water pushing outwards on the cell wall (turgor) in plant cells helps:

1   to keep **stems upright**

2   to keep **leaves flat** so they can better absorb sunlight.

The water potential inside most **animal** cells is often the same as the solution in which the cells are naturally bathed. (See Chapter 12, Kidney Function.) Thus there is little movement of water by osmosis into or out of the cell.

However, a red blood cell placed in a solution with a relatively high water potential, starts to take in water by osmosis, and since there is no cell wall to resist the increased pressure that results, the cell **bursts**.

The loss of water by osmosis

Plant cells placed in a solution of relatively low water potential, lose water from their vacuoles. As a result they lose their internal pressure since the cytoplasm is no longer being forced against the inelastic cell wall. They become soft (Figure 3.7).

When a cell loses its internal pressure, it is said to be **flaccid**.

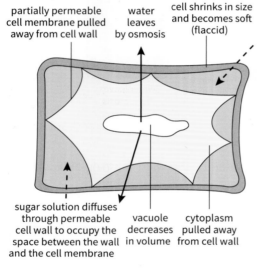

vacuole decreases in volume and pressure

water leaves vacuole by osmosis

Pressure is no longer exerted on cell walls the cell becomes flaccid and decreases in length and width.

Figure 3.7 A flaccid plant cell

Further exposure to a bathing solution of lower water potential (such as a sugar solution) will draw so much water from the vacuole that the cytoplasm is pulled away from the cell wall (Figure 3.8). (Such a condition is called **plasmolysis.**)

partially permeable cell membrane pulled away from cell wall

water leaves by osmosis

cell shrinks in size and becomes soft (flaccid)

sugar solution diffuses through permeable cell wall to occupy the space between the wall and the cell membrane

vacuole decreases in volume

cytoplasm pulled away from cell wall

Figure 3.8 A (plasmolysed) plant cell after it has been left in a sugar solution of low water potential for some time

Animal cells placed in solutions of **lower water potential**, lose their shape and what turgidity they have, as water moves out of their cytoplasm. A red blood cell shrinks in size and its cell membrane becomes unevenly creased ('**crenated**') as shown in Figure 3.9.

Animal cells placed in a solution with a **high water potential** (e.g. pure water) take in water by osmosis. They have no inelastic cell wall to resist the intake and to make them turgid, so they **burst.**

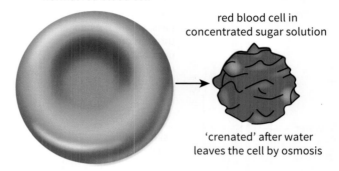

normal red blood cell

red blood cell in concentrated sugar solution

'crenated' after water leaves the cell by osmosis

Figure 3.9 A normal and a crenated red blood cell

In animals, the cells of the body are usually bathed in a fluid that is maintained at a constant water potential (one of the jobs of the kidneys, in vertebrates). This ensures that the concentration of chemicals in the cell cytoplasm remains constant and, therefore, so too is the rate of metabolism.

PRACTICAL

Investigation of the effect on plant tissues of the entry and exit of water by osmosis

Apparatus and materials:	Two large potatoes
	Two containers (e.g. beakers) at least 10 cm deep
	A sharp knife
	A supply of sucrose (table sugar) or table salt (sodium chloride)
	A ruler measuring in mm
	A supply of water

Method:

Peel one or both potatoes to provide six similar strips ('chips') at least 7 cm long, and no more than 0.5 cm × 0.5 cm in cross-section. Then cut all the strips to the same length and record that length.

Fill one beaker to within 2 cm of the top with water, and the other with a concentrated sucrose or sodium chloride solution.

Submerge three of the potato strips in each of the beakers.

Leave the strips for at least half an hour (the more dilute your sugar/salt solution, the longer you should leave them), then remove them, measure them and record their lengths. Note the texture of the strips.

Results:

The strips in the water will have increased in length and will be firm to the touch; those in the sugar or salt solution will have decreased in length and be soft to the touch.

Explanation:

For the strips in water, the cell sap in the cells of the potato will have a lower water potential than the water the strips were submerged in, thus water has entered the cells, stretching them and increasing the pressure inside them, making the potato feel firm.

For the strips in the sucrose or salt solution, the water potential of the cell sap is higher than the water potential of the solution, thus water has moved out of the cells. The cells have decreased in size and lost their firmness. Thus the strips are shorter and are soft to the touch.

The effect of the gain or loss of water on a plant

When a plant has access to water and absorbs it by osmosis, its cells become firm as the water enters the cell vacuoles and presses outwards on the cell walls. Like the potato strips, the tissues of the plant become firm and, when this happens in the stem and leaves, the plant is supported and held upright and the leaves are firm and held open to the sunlight for photosynthesis.

Loss of water from cells makes stems soft and no longer able to support the plant and leaves curl and are less able to photosynthesise.

Place a fruit sweet between your teeth and your cheek and leave it there for about a quarter of an hour without biting or sucking it. Then remove it and feel the inside of your cheek with your tongue. Write down a description of how it feels and attempt an explanation.

Progress check 3.3

1   Why, when a plant cell and an animal cell are both placed in water, does only the animal cell burst?

2   Figure 3.10 shows two liquids separated by a partially permeable membrane at the start of an experiment.

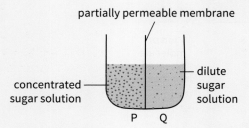

Figure 3.10

Which part of Figure 3.11 shows and explains the results 20 minutes later?

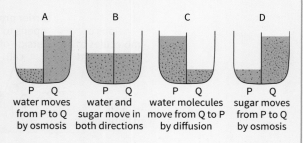

A	B	C	D
P Q	P Q	P Q	P Q
water moves from P to Q by osmosis	water and sugar move in both directions	water molecules move from Q to P by diffusion	sugar moves from P to Q by osmosis

Figure 3.11

3   Why do the leaves and stems of a plant shoot, whose cells have started to undergo plasmolysis, begin to droop?

3.04 Active transport

We have seen how chemicals move into and out of cells by diffusion down a concentration gradient. Sometimes, a living cell may be in need of a chemical, even though there is a lower concentration of it outside than inside the cell. In such a case, the chemical would have to be absorbed **against a concentration gradient**. This can be achieved only with the use of **energy** released by **respiration** inside the cell in conjunction with the cell membrane. This energy allows the cell to absorb the chemical through the cell membrane and prevent any of the chemical already inside the cell from leaving.

This process is called active transport and is defined as **the movement of particles through a cell membrane from a region of lower concentration to a region of higher concentration using energy from respiration**.

The energy released in respiration is stored in the chemical ATP in the mitochondria. Cells that undergo active transport characteristically have a large number of mitochondria.

The energy used in this process not only affects the kinetic energy of the molecules that are being absorbed, but it is used to move molecules in the cell membrane so they actively 'pump' the required molecules into the cell.

This situation arises in **plant roots** where the **ions** needed for a plant's metabolism may be in very short supply in the soil water. Ions are thus absorbed by root hair cells by active transport.

The situation also arises in the **small intestine** of an animal, when digested food (such as **glucose**) is absorbed by the epithelial cells of the **villi** by active transport.

3.05 How particles are moved by active transport across a cell membrane

Particles, usually in molecular form, are taken up by cells against a concentration gradient. This is achieved by proteins, called **carrier proteins**, that are bedded in the cell membrane (Figure 3.12). They are the same thickness as the membrane and thus they are in contact with the surrounding of the cell on the outside and the cytoplasm of the cell on the inside.

Energy is used to open up a channel on the part of the molecule facing the outside of the cell. The molecule or ion to be absorbed enters the channel and binds to a special site in the centre of the carrier protein molecule. Each carrier protein will bind to only one particular molecule. Thus, like **enzymes**, they are specific. Further energy is then used to close off the opening to the outside of the cell, and to open up a similar channel, this time into the cytoplasm of the cell. The molecule being absorbed is then released from its binding site into the cell's cytoplasm.

The carrier protein then reverses the process, again using energy so that it is available to take in another molecule.

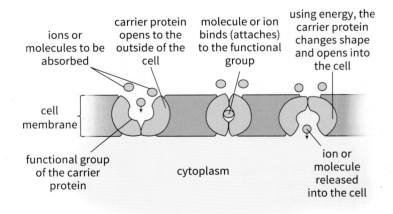

Figure 3.12 Diagram of carrier protein at work

Worked example

Describe how the structure of a cell membrane is adapted to the process of active uptake.

Cell membranes contain protein molecules called 'channel proteins'. These molecules fit loosely together leaving minute channels between them extending from the outside of the cell to the cell cytoplasm. Suggest how these channel proteins may play a part is the processes of diffusion and osmosis.

Answer

The question calls first for a realisation that there is a special structural feature of the cell membrane making it able to undergo active uptake. That feature is the presence of carrier proteins. Since they have to work against a concentration gradient (there may be a lower concentration of the chemical to be absorbed outside than inside the cell) then energy must be used. All energy within a cell is initially released by respiration and, in this case,

the energy is used to move the carrier proteins. First the carrier protein opens to the outside and allows the molecule to be absorbed to attach (bind) to the protein. Only one type of molecule will bind as the site is not suitable for any other molecule. The protein then changes shape again (again using energy) – closing the outer opening, and opening into the cytoplasm of the cell. The molecule is released into the cell cytoplasm, then changes shape again, closing to the inside and opening to the outside ready to bind with another molecule to be absorbed. (Note that the complete cycle of carrier protein movements is described as a continuous process.)

Diffusion requires pores in the membrane before molecules can enter. The channel proteins would provide the pores. However, the size of the pores might prevent larger molecules from entering. Water is a small molecule and thus could enter by osmosis, and the pores may be too small to allow larger molecules to enter – making the membrane semi-permeable.

Chapter summary

- [ ] You have learnt how cells are involved in the processes of diffusion, osmosis and active transport.

- [ ] You have learnt how these processes are important to living structures.

- [ ] You have learnt how to demonstrate these processes experimentally.

- [ ] You have also learnt about the factors that affect them.

Exam-style questions

1 Describe how different substances in a leaf move by diffusion during a 24-hour period. [6]

2 a Figure 3.13 shows a piece of partially permeable tubing, tightly tied at each end, and containing a concentrated sugar solution that is coloured with blue dye. It has been placed in a beaker of pure water.

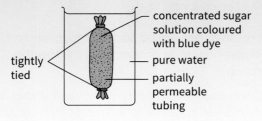

tightly tied — concentrated sugar solution coloured with blue dye
pure water
partially permeable tubing

Figure 3.13

Describe what will happen to the tubing and its contents over the next 20 minutes. [3]

b After 20 minutes, apart from what happens to the tubing, the water in the beaker has turned blue. With reference to diffusion and osmosis, explain the results of this experiment. [7]

3 a Explain how a plant root absorbs from the soil:

i) water [6]

ii) essential mineral ions that are in very short supply. [4]

b Suggest why a plant may have great difficulty in absorbing essential mineral ions that are in very short supply in a water-logged soil. [4]

The chemicals of life

4.01 Biological molecules

Three of the most common organic molecules found in living organisms are:

* **carbohydrates**

* **fats**

* **proteins**.

Carbohydrates

Carbohydrates are organic chemicals containing the elements **carbon**, **hydrogen** and **oxygen** only. The ratio of atoms of hydrogen to atoms of oxygen in a carbohydrate molecule is always 2:1.

(It may help to remember this to know that 'carbo' refers to the carbon, 'hydrate' refers to water – where the ratio of H to O is also 2:1.)

Carbohydrates with large molecules, such as **starch**, glycogen and cellulose, are **insoluble**. (Although it is common to refer to a 'starch solution', it is really a starch suspension.)

They are synthesised in living organisms by linking together molecules of simple sugar (**glucose**). Large carbohydrate molecules can be broken down into simple sugars.

Smaller carbohydrate molecules are soluble and occur as:

* 'complex' sugars, such as **maltose** and **sucrose** (table sugar), all with the formula $C_{12}H_{22}O_{11}$ or

* 'simple' sugars, such as **glucose** or **fructose**, with the formula $C_6H_{12}O_6$.

Fats

TIP

Watch out for questions that refer to oils and remember that oils are fats that are liquid at 20°C.

Fats are organic chemical substances that contain the elements **carbon**, **hydrogen** and **oxygen** only. This time, however, the ratio of H to O in the molecule is very much higher than 2:1. They are all insoluble in water and are formed by the joining of a **glycerol** molecule with **fatty acid** molecules.

Proteins

Proteins contain the elements carbon, hydrogen, oxygen and **nitrogen** (and often other elements such as sulfur and phosphorus).

They are large, usually insoluble, molecules that are built up from simple, soluble units known as amino acids, of which up to 20 are used in the production of protein in living organisms.

Figure 4.1 Amino acids linking to form a protein molecule (each different symbol represents one of the 20 different amino acids)

A relatively few amino acids link together to form a **polypeptide**, while polypeptides link together to form protein.

Amino acids are of different sizes and shapes, thus, when linked together to form protein molecules, the proteins formed will also be of different shapes.

This structural difference is very important when the protein in question is an **enzyme** since part of the molecule, known as the **active site** is of exactly the right shape to fit the molecule (the **substrate molecule**) of the chemical reaction that it will catalyse.

Similarly, if the protein is an **antibody** in the blood, part of its molecule (the **binding site**) will be exactly the right shape to stick to ('bind with') the pathogen or toxin that the antibody has been made to counteract.

Progress check 4.1

1   Name the elements found in i) carbohydrates ii) fats iii) proteins.
2   Name the units from which i) fats, ii) starch and iii) proteins are made.

Tests to show the presence of carbohydrates, fat and protein

These are often referred to as 'food tests'. The tests are shown in Table 4.1 (carbohydrates), Table 4.2 (fat) and Table 4.3 (protein).

Carbohydrates

Benedict's solution does not distinguish between the complex sugar such as **maltose**, and the simple sugar such as **glucose**. In their reaction with Benedict's solution, both sugars work as chemicals known as reducing agents, and thus belong to a group of sugars referred to as **reducing sugars** all of which react in this way.

Carbohydrate	Chemical reagent used	How test is carried out	Result
Starch	iodine solution	put a few drops on the substance to be tested	blue-black colour if starch is present brown if starch is absent
Maltose and glucose	Benedict's solution	add a few drops to a solution of the substance to be tested and heat in a water bath at 90 °C	red/orange/yellow/green if either of the sugars is present blue if neither of the sugars is present

Table 4.1 Tests for carbohydrates

Worked example

a   **Explain the chemical similarities and differences between glucose and sucrose.**

b   **Explain how you could distinguish between them by performing a simple food test.**

Answer

a   It is best to split this part into its two natural sections – similarities and differences. The first similarity is that they are both sugars and they are both soluble so that they can be more easily moved from place to place within an organism. The fact that they are sugars is indicated by the ending '-ose'. They will therefore both also be carbohydrates – a term you should explain. The clue is, again, in the word: 'carbo-' because they contain carbon and 'hydrate' because they contain the same elements as water – hydrogen and oxygen, and in the same ratio as well – 2 Hs to each O. Some carbohydrates can be very large molecules, but these sugars are comparatively small ones.

The differences are that one, glucose, is a simple sugar and is the basic unit of many larger carbohydrates (such as starch). Sucrose is a complex sugar and one molecule of sucrose ($C_{12}H_{22}O_{11}$) is approximately twice as big as one molecule of glucose ($C_6H_{12}O_6$).

b   This part hinges on the fact that the simple test for sugars identifies only simple (or 'reducing') sugars. Glucose is a reducing sugar but sucrose is neither a reducing nor a simple sugar. Therefore, the Benedict's test will identify the glucose but not the sucrose. You are asked how you would distinguish between them, so you will need to give a description of the test and state the result you would expect. The test is to heat a little of each sugar in test-tube together with a few drops of Benedict's solution. The glucose will turn red, but the sucrose will remain the colour of the Benedict's solution – blue. (NB If you mention a bunsen burner for heating, you would be advised also to mention the use of goggles.)

Fat – the ethanol emulsion test

Reagent used	How test is carried out	Result
Ethanol	a dried sample of the substance to be tested is mixed with ethanol, which is then poured into a test-tube of water	water turns cloudy if fat is present water remains clear if fat is not present

Table 4.2 Tests for fat

> **TIP**
>
> When performing ethanol emulsion test, the test is more reliable if a dried sample is used, but if it is not possible to dry the sample, then the ethanol drained from it may already be cloudy, indicating the presence of a fat.

Proteins – the biuret test

Reagent used	How test is carried out	Result
biuret solutions 1 and 2	add equal volumes of solution 1 and 2 to a solution of the substance to be tested	purple colour if protein is present blue colour if protein is absent

Table 4.3 Test for protein

Biuret 1 contains sodium hydroxide that is harmful to the skin and thus appropriate safety precautions should be taken.

Since 'biuret' is a chemical – not a person as in 'Benedict's solution' – it does not begin with a capital letter, unless it is the first word in a sentence.

>
> **TIP**
>
> Never say that a food test is 'positive' or 'negative'. State the observed colour and the conclusion you draw from that result.

The test for vitamin C

As indicated, these tests are usually referred to as 'food tests' because they are mostly carried out to investigate the presence of chemicals in various foods. Another important constituent of foods is **vitamin C** and there is a test that is used to decide whether a food contains vitamin C. The reagent used is known as **DCPIP** which is in the form of a blue solution.

Reagent	How test is carried out	Result
DCPIP	add solution of substance drop by drop to DCPIP	DCPIP changes from blue to colourless if vitamin C is present

Table 4.4 Test for vitamin C

Progress check 4.2

1 Name the chemicals used to test for:
   i) starch; ii) vitamin C.

2 A food test gives a purple colour. What reagent has been used and what chemical is present?

3 When using Benedict's solution, why is it not possible to be absolutely sure that a food contains glucose?

DNA

Chromosomes, situated in the **nuclei** of all living cells (except bacteria, which have no true nucleus) have, as their major component, the chemical substance **DNA**. The DNA molecule, looking rather like a very long, twisted rope ladder, is made up of two strands (of alternating sugar and phosphate units) held together by pairs of chemical units called **bases** (these bases form the 'rungs' of the ladder). The molecule is described as a **double helix** (because the Greek word *helix* means a 'spiral'.)

There are only **four** bases, known by their initial letters **A**, **T**, **C** and **G**, and the sequence in which they occur in one of the two DNA strands is responsible for the sequence of amino acids in the protein molecules that are made in the cell. Since the sequence of bases on a DNA molecule is likely to be different for each (sexually produced) individual, it follows that no two individuals will make protein molecules with exactly the same sequence of amino acids.

As stated previously, the bases are always paired when forming the 'rungs' of the DNA double helix, but there is a 'rule of base pairing', since **A** always pairs with **T**, and **C** always with **G**.

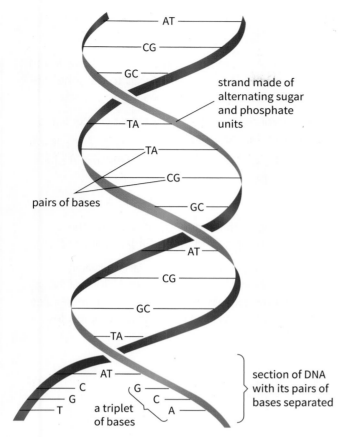

Figure 4.2 Diagram of DNA molecule

Water as a solvent

Water is often described as 'the universal solvent'. While this is not chemically true, living organisms are fundamentally reliant on water as a solvent.

- All reactions of metabolism take place in solution.

- The cytoplasm of cells is 70–90% water, and the chemicals within it, apart from storage chemicals, are in solution.

- Chemicals are absorbed by plants and animals in solution (ions from the soil in plants and ions, vitamins and digested foods from the alimentary canal in mammals).

- The enzymatic process of digestion is a type of reaction known as **hydrolysis** during which water molecules are used to break down large organic molecules into smaller ones.

- Once absorbed by an organism, water transports the absorbed chemicals in solution – ions in the xylem and sucrose and amino acids in the phloem of plants; amino acids and glucose in the blood stream of mammals

- Water is an efficient transporter of heat – an important property of the water in blood plasma.

Chapter summary

☐ You have learnt about the chemical structure of the three major organic substances, which are also major constituents of our diet.

☐ You have then learnt how the chemical tests ('food tests') for identifying these substances are carried out.

☐ Students following the Supplement material have learnt the structure of the DNA molecule.

Exam-style questions

1 a Three of the main organic chemicals found in the living organisms are formed as follows:

   P a long chain of identical molecules linked together

   Q a long chain of different molecules linked together

   R molecules of two different types only linked together

   Identify **P**, **Q** and **R**. [3]

   b i) Name the elements found in **all three** of these chemicals. [3]

      ii) Name the element that **must** always be present in **only one** of these chemicals. [1]

   c Suggest reasons for the fact that only genetically identical organisms have exactly matching protein molecules. [5]

2 a DNA is a molecule in the form of two linked strands. State the term used to describe the appearance of these two linked strands. [2]

   b Where in a cell is the greatest concentration of DNA found? [2]

      The sequence of bases on one strand of a DNA molecule is shown here.

      G-A-A-T-T-G-C-A

   c What is the sequence of bases on the matching strand in this DNA molecule? [4]

3 Starch and proteins are both large organic molecules important in the metabolism of living organisms. Explain why it is that starch molecules are all very similar, whereas protein molecules are largely all different. [5]

Enzymes

5.01 Enzymes

The rate at which chemical reactions occur can by altered by the presence of particular chemicals called catalysts. Catalysts need to be present only in **very small amounts** and they are not used up by the reaction, which means that they can go on working, continuously catalysing the same reaction, so long as the reactant molecules (the **substrate** molecules) are present. Catalysts usually **speed up** chemical reactions.

Within cells, the chemical reactions occurring in the cytoplasm are also under the control of catalysts – **biological catalysts** – correctly called **enzymes**. Without enzymes, the reactions of metabolism would not be fast enough to sustain life. In particular, respiration would not be able to supply enough energy, fast enough, to satisfy all the energy demands within the different tissues of the organism.

An **enzyme** is defined as **a protein molecule that functions as biological catalyst – that is, it speeds up a chemical reaction and is not changed by the reaction**.

All enzymes are made in the cytoplasm of all living organisms under instruction from genes on the chromosomes in the nucleus.

When a chemical reaction occurs using an enzyme, the molecule or molecules that take part in the reaction fit into the enzyme molecule similar to the way that a key fits into a lock – the enzyme being the lock. This is therefore sometimes known as the known as the '**lock and key**' mechanism of enzyme action, during which the enzyme is able either to break up a larger molecule – or, sometimes, to join together two smaller ones.

There is one particular **temperature** at which each enzyme works at its fastest rate. This is known as its **optimum temperature** and for most enzymes in a mammal's body this is around 37 °C. As the temperature slowly approaches the optimum, the rate of enzyme activity gradually speeds up. Above the optimum, the rate of activity begins to slow down, until, at around 60 °C, the enzyme is **destroyed** or **denatured**.

> **TIP**
> Never say that enzymes are 'killed' – they were never alive.

Thus, the rate of the reaction falls to zero (Figure 5.1).

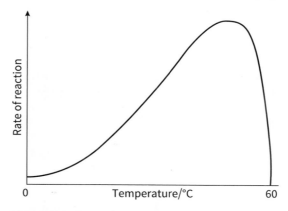

Figure 5.1 The effect of temperature on enzyme action

> **TIP**
> It may help you to understand denaturing if you remember that when an egg is cooked, the transparent part around the yolk [the albumen] turns white. If the egg is then allowed to cool down, the albumen remains white. Albumen is protein and heat has permanently denatured it.

There is one particular **pH** at which each enzyme works at its fastest rate. This is known as the **optimum pH**. For most enzymes in a mammal's body, this is around pH 7 (or slightly above), though for the enzyme **protease** in the stomach, its optimum pH is around 1.5 – created by the presence of stomach hydrochloric acid.

Either side of the optimum pH, the rate of enzyme activity gradually decreases (Figure 5.2).

NB pH is a measure of the degree of acidity or alkalinity. A pH of 7 is neutral, below 7 is acidic and above 7 is alkaline.

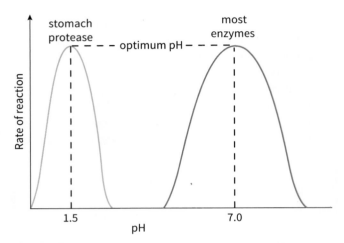

Figure 5.2 The effect of pH on the rate of an enzyme-controlled reaction

Progress check 5.1

1   What is the name of the chemical on which an enzyme acts during a chemical reaction?

2   Why do enzymes need to be present in only small quantities even when the chemicals on which they act may be in large quantities and involved in a continuous reaction?

3   Can you think of any enzymes that do not have an optimum temperature of around 37 °C?

Worked example

Explain the effect of increasing the temperature on an enzyme-controlled reaction.

At first sight, the question appears to be referring to the fact that increased temperature increases the rate of reaction. But here are three separate parts to the answer. The first is to say that there is an increases in the rate of the reaction up to the optimum temperature (i.e. the temperature at which the enzyme is working at its fastest. The second is to describe what happens immediately after the optimum temperature. A further, relatively small increase in temperature will decrease the rate until the enzyme is not active at all. After this point (the third stage), the enzyme is denatured, so an increase in temperature will have no effect, as the enzyme is permanently denatured (or destroyed). The rate of reaction remains at zero.

How enzymes work

This mechanism is often referred to as the **'lock and key hypothesis'**. A hypothesis is a suggested explanation, based on some evidence, but needing further investigation.

The job of an enzyme (e.g. those used in digestion) is often to break down a large molecule (the **substrate molecule**) into smaller molecules (the **product**).

Importantly, one particular metabolic reaction is catalysed by one particular enzyme only. The enzyme is described as being **specific** to that reaction. This specificity is explained as follows:

Each enzyme is a molecule with a **specific shape**. On part of its surface is the **active site** (the 'lock') – a section into which its substrate molecule (the '**key**') fits **exactly**. When the substrate molecule is in position in the active site (forming the '**enzyme–substrate complex**'), the enzyme then slightly stresses (or '**bends**') the substrate, splitting it into two product molecules that drift away from the enzyme molecule leaving its active site free to operate again. Often by the incorporation of a molecule of water, the newly exposed ends of the product molecules are chemically '**sealed**' so that they will not re-join. This type of enzyme-controlled reaction, common in digestion, is known as **hydrolysis**.

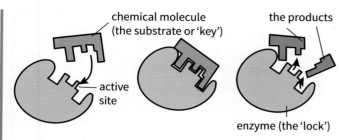

chemical molecule (the substrate or 'key')

active site

the products

enzyme (the 'lock')

Figure 5.3 The 'lock and key' explanation of enzyme action

This hypothesis explains enzyme action since:

1 Only the correct enzyme–substrate combination can work.

2 Increased temperature increases the kinetic energy of the molecules and thus the rates at which the molecules of enzyme and substrate move and collide with each other. This increases the rate with which substrate molecules enter and product molecules leave the active site of the enzyme.

3 Excessive heat causes the atoms of the enzyme molecule to move so violently that they change position relative to one another – thus changing the shape of the active site, causing the enzyme to cease functioning.

4 Changes in pH are known to alter the shape of large molecules like proteins. When that protein is an enzyme, then the more the shape of the active site is changed, the less efficiently will the enzyme work.

PRACTICAL

Make sure you know how the effect of temperature and of pH can be demonstrated.

1 The effect of temperature

Apparatus: Plastic syringe

A beaker of ice and water

2 water baths, one set at 25 °C and the other at 40 °C

3 test-tubes

3 dropping pipettes

A white tile (preferably dimpled)

Materials: A supply of 1% starch solution

A supply of 5% amylase solution (amylase is an enzyme that digests starch to reducing sugar)

Iodine solution

Method:

Using the syringe, transfer 5 cm³ of the starch solution to each of the test-tubes.

Carefully rinse the syringe (or use another clean one) and add 1 cm³ of the amylase solution to each test-tube.

Shake the tubes to mix the contents and place one in the ice/water beaker and one in each of the water baths.

At 1 minute intervals, take a drop from each test-tube with a pipette, rinsing the pipette each time before taking a sample and transfer it to the white tile. Test each drop with one or two drops of iodine solution.

Results and conclusion:

At first, all drops of iodine turn blue/black. Eventually, a drop from the test-tube incubated at a temperature nearest to the optimum temperature for the amylase will remain orange/iodine-coloured. This indicates that all the starch has been digested.

Drops from which test-tube will be most likely always to turn blue/black and why?

2 The effect of pH

The investigation is carried out in a way that is similar to the experiment on the effect of temperature, but this time, using only one water bath at 40 °C and five test-tubes labelled 1–5. Also, the following solutions are required:

M/20 sodium carbonate (an alkali)

M/10 ethanoic (acetic) acid

Place in all test-tubes: 5 cm³ starch solution, 1 cm³ amylase solution and add the following:

• To test-tube 1, 1 cm³ sodium carbonate (this will provide a pH of 9)

- To test-tube 2, 0.5 cm³ sodium carbonate (providing pH around 7.5)

- To test-tube 3, add nothing (pH 7 to very weakly acidic)

- To test-tube 4, 2 cm³ ethanoic acid (pH 6)

- To test-tube 5, 4 cm³ ethanoic acid (pH 3).

After shaking, all test-tubes are placed in the water bath and drops tested at 1 minute intervals with iodine solution as before.

NB The syringes must be washed out carefully before taking a fresh sample of drops.

Record your results.

See also the experiments in Chapter 8 using biological washing powders.

Progress check 5.2

1   What might your results tell you about the pH at which amylase works in your digestive system?

2   Which molecule in an enzyme-controlled reaction is normally referred to as 'the key'?

TIP

Remember that the names of enzymes commonly end in **'ase'** (e.g. 'amylase')

Chapter summary

☐ You have learnt that enzymes are biological catalysts.

☐ You have learnt the effect of temperature and pH on the action.

☐ You have carried out practical investigations into these effects.

☐ Students following the supplementary course have learnt how enzymes and their substrates operate like a lock and key.

Exam-style questions

1   a   Figure 5.4 shows the effect on the rate of an enzyme-controlled that normally works in the human body as the pH is raised to 7.

Copy and complete the graph to show what will happen to the rate of reaction as the pH is raised further to 9. [3]

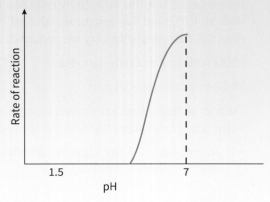

Figure 5.4

b   For the same enzyme, state what would happen to the rate of reaction at pH 7 if the temperature is gradually raised from 20 °C to 70 °C, then slowly returned again to 20 °C. [5]

2   With reference to the 'lock and key' explanation of enzyme action, explain the effects of increased temperature on an enzyme-controlled reaction. [8]

3   a   Explain the significance of the enzyme-substrate complex in an enzyme-controlled reaction. [4]

b   Explain the term hydrolysis when applied to some forms of enzyme-controlled reactions. [3]

Chapter 6

Plant nutrition

Learning outcomes

By the end of this chapter you should understand:

- ☐ The process of nutrition in plants
- ☐ How to show, by experiment, what a plant requires for photosynthesis
- ☐ How external factors affect the rate of photosynthesis
- ☐ How leaves are adapted for the process of photosynthesis

Nutrition appears at the beginning of Chapter 1 as one of the characteristics of living organisms. It involves the taking in of materials for energy, growth and development.

6.01 Plant nutrition (photosynthesis)

Green plants are described as autotrophic. They use small molecules readily available around them (**carbon dioxide** from the air and **water** from the soil) and build them up into large organic molecules (of the carbohydrate glucose) during the chemical reaction known as **photosynthesis**.

The linking together of carbon dioxide and water molecules requires energy and this is provided by sunlight, which is trapped by the green, magnesium-containing chemical called **chlorophyll**, found within the chloroplasts that are the site of photosynthesis inside a cell.

Thus **photosynthesis** is defined as **the process by which plants manufacture carbohydrates from raw materials using energy from light**.

This is shown in the following word equation:

$$\text{carbon dioxide} + \text{water} \xrightarrow[\text{light energy}]{\text{chlorophyll}} \text{glucose} + \text{oxygen}$$

$$6CO_2 + 6H_2O \xrightarrow[\text{light energy}]{\text{chlorophyll}} C_6H_{12}O_6 + 6O_2$$

TIP

Remember that the equation is the reverse of what happens in respiration; oxygen is given out and carbon dioxide is taken in. Each chemical begins with a '6' – except glucose, which begins C_6.

The **energy** supplied by the **light** and used to join the carbon dioxide and water molecules together to make glucose remains locked within the glucose molecule as **chemical energy**. When the glucose is later broken down during respiration, the energy is then released and used for various activities within the organism (see Chapter 11 on Respiration).

Progress check 6.1

1 What happens, chemically, during photosynthesis?

2 Which of the characteristics of living organisms is a plant carrying out when it photosynthesises?

The glucose manufactured during photosynthesis may be put to a number of uses in a cell:

- Energy release during respiration

- Conversion to starch (and to glycogen in fungi and animals) and used as an *energy store*. Plants first store starch in their chloroplasts, but often also in their stems and roots, which may be modified for the purpose

- Conversion into cellulose to make cell walls

- With the addition of mineral ions (always nitrates but often also sulfates, phosphates and iron), conversion into proteins

- Conversion into fats – a useful storage chemical containing twice as much energy per unit mass as carbohydrate – and a component of cell membranes.

It is possible to show experimentally that carbon dioxide, light energy and chlorophyll are necessary for the process of photosynthesis, using potted plants (e.g. *Pelargonium* – a member of the geranium family). It is difficult to show that water is necessary for photosynthesis, since the removal of water from a plant may have serious effects other than on photosynthesis.

PRACTICAL

You need to know how the conditions for photosynthesis can by demonstrated experimentally.

1 Aim: **To show that carbon dioxide is necessary for photosynthesis**

Apparatus: 2 well-watered de-starched potted plants (e.g. *Pelargonium* or *Coleus*)

Polythene bag to will fit over one of the pots, cotton (to tie the polythene bag over the pot and around the stem)

A large piece of flat glass

2 bell jars

Some petroleum jelly (such as Vaseline)

A small beaker containing concentrated sodium hydroxide (follow the safety precautions when handling)

Before the experiment, the plants should be kept in a dark place (such as a cupboard) for 24–48 hours, so that any starch present in their leaves has been used up. This is important because these types of plant immediately turn any glucose they make during photosynthesis into starch and we shall be testing the leaves after the experiment to see if any starch is present. We must therefore make sure that there is no starch present before we begin the experiment.

Method:

The experiment is set up as shown in Figure 6.1.

The two plants are left side by side in sunlight for about 8 hours, after which, a leaf is taken from each plant and tested for the presence of starch.

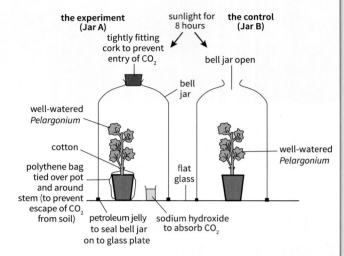

Figure 6.1 Experiment to show that carbon dioxide is necessary for photosynthesis

The starch test for a leaf:

Transfer the leaf to a beaker of boiling water for about 1 minute.

Turn off the bunsen burner if used for heating the water.

Transfer the leaf to a large test-tube which is half-full of methylated spirits.

Place the test-tube (with the leaf in the methylated spirits) into a hot water bath or to the recently boiled water and allow the heat of the water to bring the methylated spirits to the boil. As it does so, the chlorophyll is removed from the leaf but the leaf becomes brittle.

Remove the leaf from the methylated spirits, and rinse it in the hot water. This will soften the leaf.

Spread the leaf on a white tile and add iodine solution.

If starch is present the leaf will turn blue/black – if not, it will stain brown.

Observation:

If this procedure is carried out on the two leaves from Jar A and Jar B, the results should be as follows:

Leaf from Jar A stains brown. It indicates that starch is not present.

Leaf from Jar B stains blue/black. It indicates that starch is present.

Conclusion:

Since only Jar B contained carbon dioxide, but, otherwise, conditions in the two jars were identical, carbon dioxide is necessary for photosynthesis.

2   Aim: **To show that light is necessary for photosynthesis**

Apparatus: A well-watered, de-starched, potted plant (e.g. *Pelargonium* or *Coleus*)
A cork cut into two pieces
A pin

Method:

The apparatus is set up as shown in Figure 6.2.

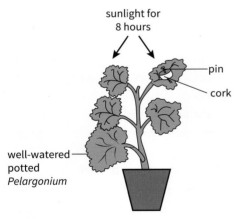

Figure 6.2 Experiment to show that light is necessary for photosynthesis

The experiment is left in sunlight for 8 hours. The cork is removed from the leaf, and the starch test is carried out on the leaf.

Observation:

Where the cork covered the leaf, the leaf stained brown. The rest of the leaf stained blue/black.

Conclusion:

Only where light had been able to reach the leaf had starch been made, thus light is necessary for photosynthesis.

3   Aim: **To show that chlorophyll is necessary for photosynthesis**

Apparatus: A well-watered, de-starched, variegated, potted plant (e.g. *Pelargonium* or *Coleus*)

A 'variegated' plant has leaves each of which has a part that is green where the chlorophyll is present. The rest of each leaf contains no chlorophyll and is often white.

Method:

Leave the plant in sunlight for 8 hours, after which remove one leaf and carry out the starch test on it.

Result:

The starch test reveals a pattern on the leaf (Figure 6.3).

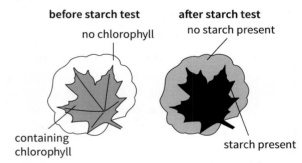

Figure 6.3 Experiment to show that chlorophyll is necessary for photosynthesis

Conclusion:

Since starch has been made only where there was chlorophyll, thus chlorophyll is necessary for photosynthesis.

The importance of a control

In all three of these experiments, a comparison is made between a leaf (or part of a leaf) that was able to function with all the requirements for photosynthesis and a leaf (or part of one) where one of the requirements is missing. In this way, the results of the experiment are shown to be valid.

The apparatus and materials that provide this comparison make up the control to the experiment. Wherever possible, biological experiments should have a **control**.

→

Experimental error

The results of any one experiment may differ from the norm. Thus, as with all biological experiments, conclusions should never be based on the results obtained from carrying out the experiment only once. In general, the more often the experiment is performed, then the more reliable are the conclusions that are drawn from the results.

After carrying out the three experiments described previously, it is possible to deduce that carbohydrate (starch) is produced when a plant has access to carbon dioxide, sunlight and chlorophyll.

There is, however, also a waste product of the photosynthesis – oxygen – and this can be shown if a water plant is allowed to photosynthesise in the laboratory.

Progress check 6.2

1   Why should the plants used in these photosynthesis experiments be place in the dark for several hours before beginning each experiment?

2   Why is there no 'control' plant in the experiment to show that chlorophyll is required for photosynthesis?

3   Figure 6.4 shows a stage in the starch test on a leaf.

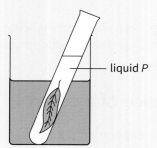

liquid *P*

Figure 6.4 Experiment to show that chlorophyll is necessary for photosynthesis

What is liquid *P*?

A   boiling water

B   cold water

C   iodine solution

D   methylated spirits

---

4   Aim: **To show that oxygen is given off during photosynthesis**

Apparatus: Large beaker

Short-stemmed funnel

Something to support the funnel (e.g. two thick coins)

Sodium hydrogen carbonate powder

Test-tube

Water weed (e.g. *Elodea* (Canadian pond weed) or *Hydrilla*)

Method:

Set up the apparatus as shown in Figure 6.4, and leave it over a 2 day period in a position where it will receive sunlight.

TIP

If you carry out this experiment, you will find it easier to position the test tube full of water over the stem of the funnel if the funnel has a short stem, the end of which is below the water level in the beaker. Fill the test tube with water, place your thumb over the end, invert the test-tube, submerge the end of the test-tube in the water, remove your thumb, then gently ease the end of the test-tube over the stem of the funnel.

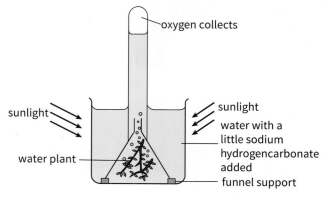

oxygen collects

sunlight

sunlight

water with a little sodium hydrogencarbonate added

water plant

funnel support

Figure 6.5 Experiment to show that oxygen is given off during photosynthesis

Result:

A gas collects at the top of the test-tube which is found to relight a glowing splint.

→

Conclusion:

Only oxygen has the power to relight a glowing splint and thus oxygen has been released during photosynthesis.

The control for this experiment is to place an exactly similar set-up in a dark cupboard for the same length of time. No gas collects in the test-tube.

Since respiration is a characteristic of living cells, then plant cells are always respiring – even when they are undergoing photosynthesis. This can be demonstrated as follows:

5   Aim: **To show that a plant that is not photosynthesising continues to respire**

Use two sets of the apparatus and materials from experiment 4 (to show that oxygen is given off during photosynthesis).

Method:

Add a few drops of hydrogencarbonate indicator to the water of one set and leave the apparatus in the light to photosynthesis.

Add the same amount of hydrogencarbonate indicator to the second set of apparatus and place it in a dark cupboard for the same period of time.

Hydrogencarbonate indicator is a pH indicator that is harmless to the water plant. It is *red* in neutral or alkaline solutions and is yellow in acidic solutions.

Results:

After about 2 days, examine the two sets of apparatus.

The water in the set left in the sunlight will still be red while the water of the set in the dark cupboard will have turned yellow.

Explanation:

The plant in the dark has not photosynthesised but has continued to respire, releasing carbon dioxide into the water. The dissolved carbon dioxide turns the water slightly acidic, causing the indicator to turn from red to yellow.

Worked example

How could it be demonstrated that oxygen is given off during photosynthesis?

Answer

Some plant material is put under an inverted funnel in a beaker of water. A test-tube full of water is placed over the end of the funnel. The experiment is left for 10 minutes and the gas that collects in the test-tube is collected and tested for oxygen. When it is shown that the gas is oxygen. this proves that oxygen is given off during photosynthesis.

This is an example of a poor, superficial answer. There should be mention of the fact that the plant is preferably a water plant. The water should have plenty of carbon dioxide for the plant to use in photosynthesis and therefore hydrogencarbonate should be added to the water to provide the carbon dioxide. It would be better to state that the apparatus is placed in light and for a good deal longer than 10 minutes. If a couple of bench lamps are used, often, enough oxygen to be tested is not obtained for a couple of days. The test for oxygen should be described (using a glowing splint that re-lights – not a lighted splint).

The release of oxygen by a water plant can be used for showing what effect varying the individual requirements for photosynthesis has on the rate at which the process takes place.

6   Aim: **To show the effect of varying light intensity on the rate of photosynthesis**

Apparatus: A large test-tube

A bench lamp

A length of water-plant stem

A timer

Method:

The apparatus is set up as shown in Figure 6.6.

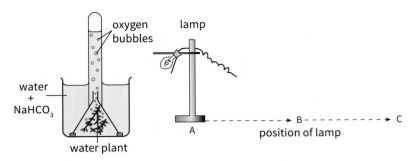

Figure 6.6 Experiment to show the effect of varying light intensity on photosynthesis

As the water plant photosynthesises, bubbles are released from the cut end of the stem. Place the bench lamp at position A. Leave the plant to acclimatise for about 10 minutes, then count the number of bubbles released by the plant for a measured period of time. Move the lamp to position B. Again, leave the plant for 10 minutes before counting the bubbles for the same measured period of time.

The lamp is moved to position C, and the procedure repeated.

NB For this exercise, and also for Exercises 7 and 8 that follow, if the bubbles of oxygen are released too quickly to count them, then the period of time that the experiment is exposed to the altered condition can be increased until the volume of oxygen released into the test-tube can be easily measured using a ruler.

It is better to carry out the experiment in a darkened room so that the only source of light reaching the plant is the bench lamp.

Results:

As the lamp is moved further away, the intensity of light reaching the plant decreases, and so does the number of bubbles/volume released per unit time by the cut stem.

Conclusion:

The rate of photosynthesis decreases with decreased light intensity.

7 Aim: **To show the effect of varying the carbon dioxide concentration on the rate of photosynthesis**

Apparatus: A large test-tube

A bench lamp

A length of water-plant stem

Some sodium hydrogencarbonate powder

Method:

When sodium hydrogencarbonate dissolves in water, it supplies the water with carbon dioxide.

$$NaHCO_3 \rightarrow NaOH + CO_2$$

The length of water-plant stem is first submerged in tap water and left for about 20 minutes to acclimatise, illuminated by the bench lamp placed about 25 cm away. With the tap water at room temperature, the number of bubbles released by the cut stem is counted over a period of, say, 3 minutes and recorded.

Next, 0.25 g of sodium hydrogencarbonate is carefully weighed out and added to the water in the test-tube. The tube is gently shaken to dissolve the powder, left for 20 minutes to acclimatise and the procedure carried out exactly as before.

This is repeated for two further additions of 0.25 g of hydrogencarbonate.

Results:

It will be noted that, as more carbon dioxide becomes available, so the number of bubbles released by the cut stem in 3 minutes increases.

Conclusion:

With increased availability of carbon dioxide, the amount of carbon dioxide released, and thus the rate of photosynthesis, increases.

8 Aim: **To show the effect of varying the temperature on the rate of photosynthesis**

Apparatus: A large test-tube

A bench lamp

A length of water-plant stem

A large beaker

A timer

A thermometer

Method:

With the apparatus set up as in the two previous experiments, the test-tube is immersed in a large beaker of water. This acts as a water bath the temperature of which may be altered using ice or warm water (e.g. from a tap). With only the temperature of the water being altered, and with the plant being left for 20–30 minutes at each new temperature to acclimatise before taking readings, the effect of varying temperature on the number of bubbles released by the cut stem is investigated.

At temperatures above room temperature, the thermometer must be checked regularly during the acclimatisation period, and the water bath topped up with warm water as necessary to maintain a steady temperature.

Result:

The higher the temperature (up to around 45 °C), the more bubbles are released per unit of time.

Conclusion:

The higher the temperature, up to about 45 °C, the faster the rate of photosynthesis.

Experimental error and the need to repeat experiments:

To reduce experimental error, each of these experiments should ideally be repeated several times. An average of the results should then be calculated.

The concept of limiting factors

Most metabolic processes are influenced by a collection of separate factors. Each factor has a particular level (the optimum) that is most effective for the process concerned. However, it is often the case that one of those factors is in limited supply and even though all other factors are present in abundance, the process is slowed by the scarcity of the one factor. This factor is then described as the limiting factor for the process.

A **limiting factor** is defined as **something present in the environment in such short supply that it restricts life processes**.

Limiting factors in photosynthesis

The availability of light, carbon dioxide, water and a suitable temperature all affect the rate of photosynthesis. However, the rate of photosynthesis in a plant at any time is governed by whichever one of these factors is in shortest supply – that is, the factor that limits the rate of photosynthesis, even though all other factors may be ideal (or 'optimum'). It is thus described as the limiting factor.

Growing plants in glasshouse ('greenhouse') systems

In colder climates, in order to obtain early and high yields of commercial crops, plants are grown in very large buildings in which the climate can be artificially controlled. They are watered regularly with appropriate nutrient solutions and the temperature, intensity and length of light per day, as well as the concentration of carbon dioxide in the atmosphere, are carefully regulated – each at its optimum level so that none of these factors becomes a limiting factor for the crop being grown. Such buildings often have glass sides, but, if artificially lit, daylight is not essential (though cheaper).

Since the atmosphere contains only 0.04% of carbon dioxide, the concentration of this gas is often raised artificially within the buildings being used to grow crops. The process is described as **carbon dioxide enrichment**.

Produce grown in this way tends to be considerably more expensive. Indeed, expense itself can become a limiting factor!

How plants obtain the necessary raw materials and carry out the process of photosynthesis

For the majority of plants, **roots** absorb the water necessary for photosynthesis from the soil, from where it has to be carried to the leaves.

See Chapter 8 for the section on transpiration.

Leaves:

1 absorb the carbon dioxide from the air

2 absorb sunlight energy which is trapped by the chlorophyll

3 manufacture carbohydrate, and

4 release the waste product and oxygen.

Sunlight provides the energy necessary to combine carbon dioxide and water molecules to form carbohydrate. That light energy becomes 'locked away' within the carbohydrate molecule as chemical energy.

Photosynthesis is thus a process in which **light energy is converted into chemical energy**. If a living organism then dismantles (i.e. breaks down) the carbohydrate molecule (during a metabolic process), the energy is released.

Until the molecule is broken down, it is often stored. The first carbohydrate made during photosynthesis is glucose, but this is soluble, affecting concentration and thus enzyme action and osmosis within a cell. It is therefore usually converted to the insoluble carbohydrate, starch.

Think of the energy you would need to pull together and tie two elastic bands fixed on hooks facing one another about 30 cm apart. Then imagine cutting the elastic bands when tied and the energy that is released if you were to do so.

Starch is first stored in the chloroplasts within the cells that are photosynthesising.

It is then converted to **sucrose** to be carried to the storage organs of a plant (see the section on translocation) where, on arrival, it is reconverted to starch, for example, in the tubers of a potato plant.

How plants absorb water from the soil

A few millimetres behind the tip of every root, and extending also for a few millimetres, lies the region of root hairs. The very numerous root hair cells thus provide a very large surface area for the uptake of water (and of ions) from the soil. The earlier section on **osmosis** explains how water is absorbed by the root hair cells.

How plants obtain carbon dioxide from the air

Most of a plant's photosynthesis takes place in the leaves (except for some specially adapted plants such as cacti). Leaves are organs containing several different tissues with cells adapted to perform their own particular roles as efficiently as possible.

6.02 Leaf structure

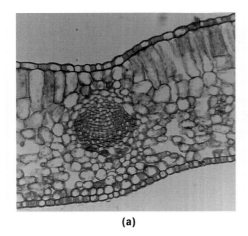

(a)

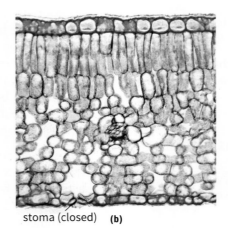

stoma (closed) (b)

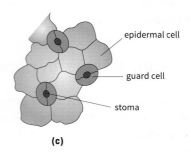

epidermal cell

guard cell

stoma

(c)

Figure 6.7 (a) Transverse section of a dicotyledonous leaf: whole leaf. (b) Transverse section of a dicotyledonous leaf: stoma and sub-stomal air chamber. (c) Stomata on the under surface of a leaf

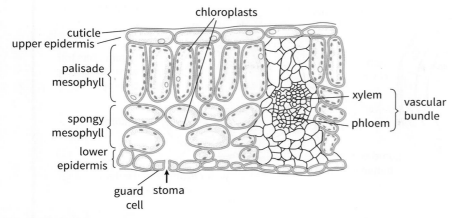

Figure 6.8 Detailed structure of a leaf of a dicotyledonous plant in transverse section

How a leaf is adapted for photosynthesis

The cuticle

The cuticle is a waxy non-cellular covering to the leaf. It is:

1   waterproof

2   transparent to allow light to enter, and

3   provides some protection from mechanical forces.

The upper epidermis

The upper epidermis is a single layer of cells that:

1   secretes the cuticle, and

2   has no chloroplasts thus allowing light to reach the cells beneath.

Palisade cells

Palisade cells are a layer (one or two, depending on species) of closely-packed, elongated cells with a water film coating their walls. They:

1   are first to receive light and

2   have the greatest concentration of chloroplasts.

Chloroplasts

Chloroplasts:

1   contain chlorophyll for photosynthesis

2   provide a large surface area for the uptake of carbon dioxide

3   are where photosynthesis occurs in the cell

4   usually convert glucose to starch and store it temporarily, and

5   in dim light, may move nearer to the illuminated surface of the cell.

Spongy cells

Spongy cells are cells with fewer chloroplasts. They are loosely packed with many air (intercellular) spaces between them. Their walls are again coated with a water film. They:

1   carry out some photosynthesis

2   allow gases to freely diffuse throughout the leaf.

Palisade and spongy cells are the photosynthesising cells in the middle of a leaf. They are often referred to as mesophyll cells. Together, their walls provide an enormous surface area for gaseous exchange as they absorb carbon dioxide and release oxygen during photosynthesis.

The vein (vascular bundle)

The vein (**vascular bundle**) contains two types of tissue, the xylem and the phloem.

The xylem:

1   strengthens the leaf helping to resist tearing

2   supports the leaf, holding it out flat so as to receive maximum sunlight, and

3   transports water and ions to the leaf.

The phloem:

1   transports dissolved sugar (sucrose) mostly away from the leaf and

2   transports amino acids both from and to the leaf as required.

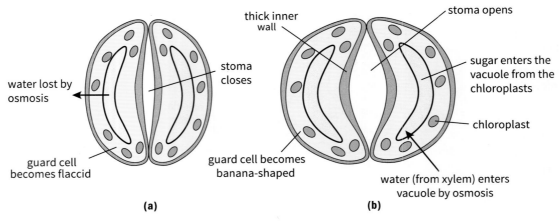

Figure 6.9 How guard cells work

The lower epidermis

The lower epidermis may be coated with a thin waxy cuticle. It forms the lower protective boundary to the leaf. The cells do not contain chloroplasts, except for the guard cells that are specialised epidermal cells. Guard cells occur in pairs, each pair controlling the opening and closing of the pore or **stoma** (plural: stomata) between them.

The function of a stoma during gaseous exchange:

1 to allow carbon dioxide to enter the leaf for photosynthesis

2 to allow oxygen to leave the leaf during photosynthesis

3 to allow some **exchange of these gases** in the opposite direction in low light intensities when the plant is respiring faster than it is photosynthesising.

Progress check 6.3

1 How does most of the carbon dioxide reach the palisade cells of a leaf?

  A   through the cuticle

  B   through the intercellular spaces

  C   through the phloem

  D   through the spongy mesophyll cells

2 How does water for photosynthesis reach the leaf?

TIP

Remember **not** to say that **water** passes out of a leaf. Always say 'water vapour' (which is a gas) passes out of the leaf by diffusion.

Stomata also allow water vapour to leave a leaf during transpiration. Guard cells control the stoma as a result of their degree of turgidity. (See osmosis.)

TIP

The guard cells become turgid, but the thicker inner walls do not stretch as much as the outer thinner one causing each cell to 'banana' creating a gap (a stoma) between them.

How a leaf is involved in the process of photosynthesis

1 Carbon dioxide diffuses down a concentration gradient from the atmosphere through the stomata into a leaf.

2 Carbon dioxide diffuses freely throughout the leaf in the intercellular spaces.

3 Carbon dioxide dissolves in the film of water that surrounds the mesophyll cells and that has been delivered to the leaf in the xylem of the vascular bundles.

4 Carbon dioxide diffuses in solution into the mesophyll cells and passes to the chloroplasts where photosynthesis occurs.

5 Sugar made by photosynthesis is carried away (translocated) from the leaf in the phloem of the vascular bundles.

6 Oxygen diffuses from the mesophyll cells into the intercellular spaces and out through the stomata, across a concentration gradient into the atmosphere.

The importance of photosynthesis to the living universe

Photosynthesis produces the carbohydrate with its stored chemical energy. Almost all other forms of life rely on carbohydrate, although plants may convert this carbohydrate into protein or fat before it is passed on.

The oxygen produced by photosynthesis is essential for the respiration of most life forms. The process of photosynthesis uses up the carbon dioxide released by respiration, and converts it into carbohydrate. (Also, see the carbon cycle.)

6.03 Mineral nutrition in a plant

The importance of nitrate ions

Living organisms need proteins for growth and repair. Plants have to manufacture (**synthesise**) their own **proteins** and this they do by converting their carbohydrates first into amino acids and then linking the amino acids together to form proteins.

The additional element necessary to convert carbohydrate into amino acid is **nitrogen** and although the atmosphere is 79% nitrogen, plants cannot make direct use of it. The nitrogen used by a plant is absorbed from the soil as the **nitrate ion** (NO_3^-) via the root hair.

The importance of magnesium ions

As with nitrates and all other ions, **magnesium ions** are also absorbed from the soil through the root hair. Magnesium is the central atom in a chlorophyll molecule. Plants lacking in magnesium ions are unable to make chlorophyll and thus their leaves are yellow in colour (a condition known as **chlorosis**).

PRACTICAL

A demonstration of the effects of a lack of nitrates and of magnesium ions in a growing plant

Apparatus: 2 small cuttings or seedlings (e.g. sorghum)

2 containers

Cotton wool

Black paper or black polythene

Culture solutions

Method:

Two seedlings or small cuttings, with the same number of leaves, are selected from a quick-growing plant and held in the top of two containers (A and B) using cotton wool as shown in Figure 6.10.

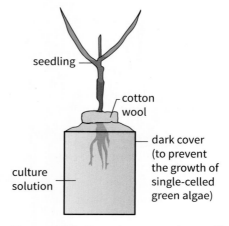

seedling

cotton wool

dark cover (to prevent the growth of single-celled green algae)

culture solution

Figure 6.10 Experiment to show effects of a lack of nitrates and of magnesium ions in a growing plant

Container A contains a complete culture solution – that is, one comprising all necessary mineral ions for healthy growth, dissolved in pure water.

Container B contains a similar solution but lacking in nitrates. The two containers are left in light at a suitable temperature (air should be bubbled into the water from time to time to supply oxygen to the roots).

Results:

Container A (the control) – the seedling grows tall and healthy with vigorous root growth.

Container B – the seedling fails to grow; the leaves begin to die and the root system fails to develop.

Conclusion:

Nitrogen, present in nitrates, is needed for the healthy growth of the seedlings.

The experiment can be repeated, but with the culture solution B lacking in magnesium ions. The results in B will be yellow leaves and only very limited growth.

Conclusion:

Magnesium is required for the manufacture of chlorophyll.

Chapter summary

■ You have learnt how a plant undergoes its nutrition using the process of photosynthesis.

■ You have also seen how the need for each condition for photosynthesis can be demonstrated experimentally.

■ You have carried out a starch test on a leaf and know the importance of each step in the test.

■ You have learnt how leaf structure is related to its function.

Exam-style questions

1 Figure 6.11 shows a section through a leaf.

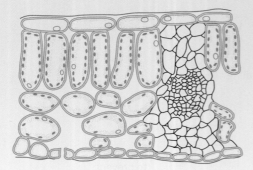

Figure 6.11

a i) Using Figure 6.11, state three different types of cell that carry out photosynthesis. [3]

ii) State the equation for photosynthesis. [3]

b Describe an experiment to show that a leaf that is green and white in colour photosynthesises only in those parts of the leaf that are green. [9]

2 a Using Figure 6.11, explain the results of the experiment in 1 b. [4]

b Explain the fact that stomata are only fully open when the plant is in the light. [9]

→

3 Figure 6.12 shows the effect of carbon dioxide concentration on the rate of photosynthesis of a plant at two different temperatures.

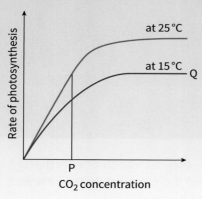

Figure 6.12

a Name the limiting factor up to the point P on the graph and explain your answer. [2]

b State the limiting factor for curve Q after point P on the graph and give a reason for your answer. [2]

c State another limiting factor for photosynthesis and explain how this factor could affect the graphs in Figure 6.12. [2]

Animal nutrition

7.01 Diet

A diet that provides all the necessary constituents in their required amounts for the good health of a person is known as a **balanced diet**. Balanced diets differ depending on the life-style, age and sex of a person (Table 7.1).

Type of person	Special requirements	Reason
child	protein carbohydrate calcium	for growth for energy for bones and teeth
active adult	carbohydrate protein	for energy to build muscles (if the job requires)
pregnant woman	iron salts calcium protein	for blood for baby's bones for making baby's cells
a woman who is breast-feeding	a balanced diet, (proteins and carbohydrates are important, also vitamin D and calcium)	for baby's growth for bone development

Table 7.1 Balanced diets

The (basal) metabolic rate of males is higher than that of females. This means that the rate at which chemical reactions occur, and thus the amount of energy released, is set at a higher level in a man's cells than in a woman's cells. Thus, for a man and a woman both involved in identical activities, the man will require more energy from his diet. Fats provide the greatest amount of energy per unit mass – just over twice as much as carbohydrates and proteins. For this reason, eating a diet rich in fats is likely to lead to obesity (being beyond overweight), when the energy consumed exceeds the energy expended by the individual.

The problems of an unbalanced diet

A person's diet may be unsuitable for healthy growth for two main reasons:

1 The balance of constituents is incorrect – leading to malnutrition.

2 There is insufficient quantity – leading to starvation.

TIP Make sure that you know and can state the difference between malnutrition and starvation.

Malnutrition

Malnutrition has many effects, of which constipation, coronary heart disease and obesity are three of the most important.

Constipation is a result of insufficient fibre in the diet (see section on nutrients). Diet lacking in fibre may, over several years, lead to bowel cancer.

Coronary heart disease can occur when animal fats and cholesterol form **atheroma** on the walls of the **coronary artery** (carrying blood to the heart muscle). This restricts blood flow in the artery, decreasing oxygen supply to the heart muscle. In severe cases, the artery may become blocked, leading to a cardiac arrest (heart attack). The risk of heart disease is increased when blood pressure is high and since atheroma decreases the bore of the arteries, it tends to increase blood pressure.

Obesity (above overweight) is associated with high blood pressure and heart disease, since it is often the result of eating too much animal fat and the heart has to work harder to move the increased bulk. It also may lead to diabetes, stress on joints and social rejection.

Starvation

Starvation results in very restricted growth and development, particularly of muscles, leading to weakness. Resistance to disease is severely reduced and death eventually follows.

Diet lacking in vitamin C

The person will develop bleeding gums and wounds do not heal properly. The disease is called scurvy. In extreme cases, anaemia (lack of iron in blood) and heart failure may occur.

Diet lacking in vitamin D

In children, growing bones become soft causing either bow legs or knock knees. The disease is called rickets. In older people, bones have a tendency to fracture.

This is because the uptake of calcium ions from the blood that are necessary for the manufacture of hard bone cannot take place without the presence of vitamin D.

Diet lacking in calcium

The skeleton does not form properly, with symptoms similar to those for lack of Vitamin D. Growth is **stunted**. In older people, bones become **brittle**.

Diet lacking in iron

Insufficient hemoglobin is made for the carriage of oxygen. The person suffers from **anaemia**. They will be lacking in energy and in severe cases, the deficiency leads to coma and death.

Hemoglobin is a protein that is built up around central 'hem' (iron-containing) groups. A lack of iron in the diet leads to a deficiency in the production of hem groups (iron deficiency) and thus the red blood cells are unable to carry sufficient oxygen to fully satisfy the needs of the

tissues. Their metabolic rate falls and insufficient energy is released.

Protein-energy malnutrition

There are two types of malnutrition that are described as protein-energy malnutrition.

- **kwashiorkor**

Kwashiorkor means 'the disease the first baby gets when the second one comes along' (from the Ga language of Ghana).

Kwashiorkor is characterised by a swollen abdomen and is caused by a protein deficiency in the diet (the baby may have been receiving all the necessary protein when being breast-fed, but may now have been weaned to a largely carbohydrate-based diet). Fluid (water that leaves blood capillaries as tissue fluid) should be reabsorbed back into the blood system. The lack of proteins in the blood raises its water potential so that water no longer passes back into the blood by osmosis, causing the abdomen to swell. The swelling is increased by the fact that kwashiorkor also causes a swollen liver.

The immune system is also affected, as the person with kwashiorkor has difficulty in producing antibodies following vaccinations for diseases such as diphtheria and typhoid.

- **marasmus**

This is caused be a deficiency not only in protein, but also in carbohydrate. The diet simply fails to provide enough protein for the development, or maintenance of muscles, or enough carbohydrates for the energy requirements of the body (NB proteins also provide energy). The patient appears extremely thin, with the outlines of ribs and other bones clearly visible beneath the skin.

The classes of food described here are from of seven important constituents of our diet. These important constituents of our diet are:

1 Carbohydrates

2 Fats

3 Proteins

4 Vitamins

5 Mineral salts

6 Fibre (or roughage)

7 Water

The principal sources of important constituents of our diet and their importance are shown in Table 7.2.

Constituent		Source	Importance
carbohydrates	starch	potatoes beans and peas yams cassava cereals	carbohydrate molecules contain energy which is released in cells as the molecules are broken down during respiration
	sugars	honey cane fruits	
fats		fat meat dairy foods egg yolk nuts avocado	fat molecules contain more energy than carbohydrates and are stored in the skin and around the kidneys
proteins		meat peas and beans fish egg white peanuts	proteins are used for growth and repair muscle is largely protein
Vitamins			
*C (ascorbic acid)		citrus fruits (fresh) cabbage	for healthy gums and skin repair
*D (calciferol)		fish liver egg yolk the action of sun on the skin	for the uptake of calcium from the gut and for strong teeth and bone formation
Mineral salts			
*calcium		flour milk	for healthy bones and teeth, for muscle action and blood clotting
*iron		liver red meat spinach	for hemoglobin – the oxygen-carrying pigment in red blood cells
fibre		fruit vegetables nuts	forms bulk in the intestines, giving the muscles of peristalsis something to push against – preventing constipation
water		drinks all foods	water is the solvent for chemicals in the body – it is the medium for all chemical reactions. It is used to cool the body and makes up about 68% of body mass

Table 7.2 The constituents of our diet

*If any of these four constituents are lacking in the diet, then the person will suffer from a **deficiency disease**.

7.02 The human alimentary canal

Human nutrition involves five basic stages:

1 **Ingestion:** The taking in of substances (such as food and drink) into the body through the mouth.

2 **Digestion:** The break-down of large, insoluble food molecules into small water-soluble molecules using mechanical and chemical processes:

   • **mechanical** – chopping and grinding food with teeth and muscular churning of the food as in the stomach, resulting in the breakdown of food into smaller pieces without chemical change to the food molecules

   • **chemical** – the breakdown of large insoluble molecules into small, soluble ones (using enzymes).

3 **Absorption:** The movement of small food molecules and ions through the wall of the intestine into the blood.

4 **Assimilation:** The movement of digested food molecules into the cells of the body where they are used, becoming part of the cells.

5 **Egestion:** The passing out of food that has not been digested or absorbed, as faeces, through the anus.

Any food that cannot be digested or absorbed is passed out of the alimentary canal during egestion. The removal of undigested food material from the alimentary canal is not a form of excretion.

The main regions of the human alimentary canal are shown in Figure 7.1.

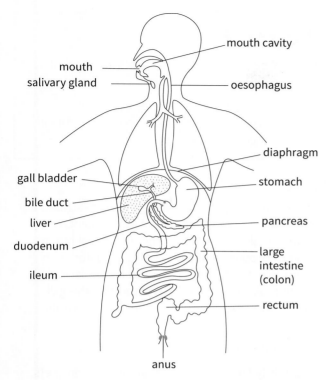

Figure 7.1 The human alimentary canal and associated organs

The functions of the main regions of the alimentary canal

The mouth

This is the hole through which food is ingested into the buccal cavity (or mouth cavity). The buccal cavity processes the food using three methods:

1 **Teeth** mechanically digest the food (see later).

2 **Salivary glands** secrete a solution (saliva) of the enzyme amylase to digest starch and the protein mucin, which is sticky, to bind the food together and to lubricate it.

3 The **tongue** rolls the food into balls (boli – bolus in the singular) and pushes them to the back of the cavity for swallowing.

TIP

It is common, and acceptable, to refer to the buccal cavity as 'the mouth', but it is preferable to use the term 'mouth cavity'.

The oesophagus

This is a muscular tube down which waves of muscular contractions (peristalsis) pass, pushing each bolus towards the stomach.

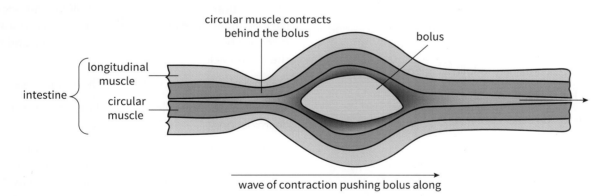

Figure 7.2 Peristalsis

The stomach

This is a muscular bag that churns the food for up to around 4 hours. Its wall secretes gastric juice which contains:

- the enzyme protease for starting the digestion of protein, changing them to polypeptides

- **hydrochloric acid** to provide the correct pH for protease to work, and to kill potentially harmful bacteria in food

- (in children only) the enzyme **rennin** to clot protein in milk.

After treatment in the stomach, the food is in the form of soup-like 'chyme', which passes through a ring of muscle (the pylorus or pyloric sphincter) that relaxes to allow the food to enter into the duodenum.

TIP Many people refer to the abdomen as 'the stomach'. This is inaccurate and the term should be used only for the muscular bag lying at the bottom of the oesophagus.

The duodenum

This is a tube about 30 cm long. It:

- receives bile via the bile duct, from the liver

- receives pancreatic juice via the pancreatic duct from the pancreas

- releases from its walls a digestive juice called intestinal juice containing:

  - the enzyme **protease**, which changes polypeptides to amino acids, and

  - the enzyme **lipase**, for fat digestion.

The pancreas

Lying between the stomach and the duodenum, the pancreas is an organ that secretes pancreatic juice, which it passes to the duodenum to help in digestion.

(The pancreas also secretes the **hormone** insulin – see Chapter 13.)

Pancreatic juice contains the following enzymes:

- **amylase** for digesting starch

- **protease** for digestion protein

- **lipase** for digesting fat.

The **amylase** completes the job started by the amylase in the buccal cavity, changing any remaining **starch** to **maltose** sugar.

The **protease**, now working in very slightly alkaline conditions, changes any remaining **proteins** to **polypeptides**. (Alkaline bile from the liver neutralises the acidity of the hydrochloric acid.)

Fats are broken up into small droplets – **emulsified** – by the bile, greatly increasing their surface area. **Lipase** can now work on them far more quickly, changing **fats** into **fatty acids** and **glycerol**.

Progress check 7.2

1 What is the difference between chemical and mechanical digestion?

2 Name the chemicals involved and the location of the only digestion that occurs in strongly acidic conditions.

The liver

The liver is the largest internal organ. It is described as the chemical factory of the body. Like the pancreas, it is not technically part of the alimentary canal but its function of producing **bile** is closely associated with digestion. Bile is a greenish-coloured fluid. It is alkaline due to the salts it contains and is stored in the **gall bladder** before it is poured into the duodenum. Its functions are:

- to **neutralise acidic chyme** from the stomach

- to **emulsify fat** (i.e. break the fat into tiny droplets – increasing the surface area on which enzymes can operate).

Large food molecules are almost ready for absorption, but first, maltose must be changed to glucose (by an enzyme in intestinal juice) and polypeptides must be changed into amino acids (by the protease in intestinal juice).

Deamination is the **removal of the nitrogen-containing part of amino acids in excess to our needs to form I followed by release of energy from the remainder of the amino acid**.

Deamination is an important function of the liver.

Two separate molecules are produced as a result of deamination:

1 **Urea**, a nitrogenous waste product (i.e. one which contains nitrogen) that passes from the liver, through the blood and to the kidneys for excretion in **urine**

2 a carbohydrate that can be changed to glycogen and stored.

Another major function of the liver is **detoxification** (the removal and breakdown of poisons – i.e. **toxins** from the blood). One such toxin is alcohol. Although the liver is able to remove small quantities of alcohol, even on a regular basis, frequent high levels of alcohol in the blood can eventually lead to liver disease (cirrhosis).

The ileum and absorption

This is the region where digested food is **absorbed** but much of the water consumed is also absorbed here. To increase its efficiency, its surface area is increased in the following ways:

- It is about 7 metres long (in a person).
- Its walls are folded ('pleated') longitudinally.
- Its walls have the appearance of velvet due to millions of microscopic finger-like projections called villi.

The term **small intestine** is often used to include the **duodenum** and **ileum**.

The colon

The walls of the colon are folded (transversely, this time) to increase its surface area for water absorption.

Infections of the colon lead to **diarrhoea** (watery faeces) – partly because water is not being absorbed, but also because water is being drawn from the cells in the wall of the colon which might help to 'flush-out' the pathogens. Treatment for diarrhoea is known as **oral rehydration therapy (ORT)** and consists of drinking a solution of mineral ions (salts) to replace the water that is lost (and thus avoid **dehydration**), but also to replace the mineral ions (salts) that are drawn from the walls of the intestine.

Food is moved steadily along the duodenum, ileum and colon by peristalsis (as in the oesophagus). The indigestible fibre forms the bulk against which the muscles of the intestines can push.

Cholera

Most bacteria that are swallowed with food are unable to survive the acidic conditions and the protease enzyme present in the stomach. However, a few, including the bacterium that causes cholera, are able shut down enough of their metabolism to survive until they reach the small intestine, where they become active again. Their activity causes considerable quantities of water and salts to be removed from the walls of the intestines, causing the patient to pass large quantities of very watery faeces.

The bacterium that causes cholera is *Vibrio cholera* and is carried in drinking water that becomes contaminated with human faeces. In areas where sewage is allowed to flow into the sea, then seafood can also be contaminated with the bacterium.

When in the small intestine, the metabolism of the bacteria produces a toxic chemical that draws ions, particularly **sodium** and **chloride** (but also potassium) out of the cells and capillaries in the intestine wall into the lumen of the gut. The effect of this is to lower the water potential of the gut contents and thus draw large volumes of water by **osmosis** out of the gut wall into the intestine. This gives rise to the high volume of watery faeces (diarrhoea), which contains a high concentration of mineral ions. Thus, ORT is important not only to replace the lost water but the lost ions as well.

The human being, like most other mammals, may be regarded as an island of water living on dry land. It is always fighting a battle against dehydration. Most of the water we consume is absorbed in the ileum along with the mineral ions, vitamins and digested foods (5–10 dm^3 per day). In order to absorb the very last drop of water but still leave enough for the comfortable removal of the faeces, the colon absorbs a final 0.3–0.5 dm^3 of water per day.

The rectum

This is a muscular storage chamber where the undigested food (faeces) is held and moulded before being pushed out through the anus during egestion.

The anus

This is the exit to the alimentary canal. It is closed by a ring of muscle (the anal sphincter), which is relaxed during egestion (or defecation).

The term large intestine is often used to include the colon, rectum and anus.

Progress check 7.3

1 Write a list of the parts of the alimentary canal in sequence.

2 How does the structure of the ileum provide it with a large surface area and for what purpose does it have a large surface area?

3 Which of these form part of the large intestine?

A colon and rectum

B duodenum and ileum

C duodenum and rectum

D ileum and colon

7.03 Teeth and the part they play in mechanical digestion

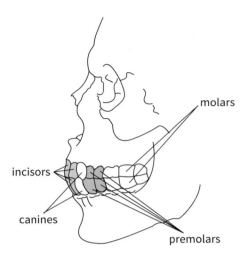

Figure 7.3 Human teeth

There are two sets of teeth in a person's lifetime. The first set (or **milk teeth**) last for around 5–14 years before they are pushed out by the permanent teeth.

NB The molar teeth are part of the permanent set only – there are no 'milk molars'.

In the permanent set, there are four types of tooth (Figure 7.3):

1   **Incisors:** There are two incisors in the front of each quarter-jaw. They are sharp spade-like teeth for biting and cutting food. They are single-rooted teeth.

2   **Canines:** There is one canine per quarter-jaw and, although conical in shape, is sharp and again used for biting and cutting food. They are single-rooted teeth.

3   **Pre-molars:** There are two premolars per quarter-jaw. The surface of each tooth has two projections (cusps), which are used for crushing and grinding food. They are double-rooted teeth.

4   **Molars:** There are three molars per quarter-jaw (though the third molars – wisdom teeth – do not usually appear until the person is at least 17 years old). The tooth surface is square with four cusps again for crushing and grinding food. They are double-rooted teeth.

Mechanical digestion carried out by teeth is therefore biting, cutting, crushing and grinding (loosely called chewing).

Tooth structure

The structure of a tooth is shown in Figure 7.4.

*   **Enamel** – This is the hardest substance in the human body and covers that part of the tooth (the crown) that appears above the gum. It is non-living and provides the surface that processes the food.

*   **Dentine** – This is a living part of the tooth, similar to bone, that supports the enamel coating to the tooth.

*   **Pulp cavity** – This is the central part of the tooth and contains nerves and blood vessels.

*   **Cement** – This is a soft bone-like substance to which ligaments are attached that hold the tooth firmly is its socket – while allowing for very limited movement (to prevent damage when chewing very hard materials).

*   **The gums** – These are soft tissue that surrounds and protects the jawbone and the roots of the teeth.

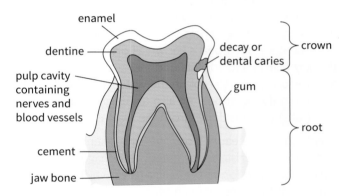

Figure 7.4 The structure of a (molar) tooth

The importance of proper tooth care

If teeth are not properly cared for, they may suffer from dental caries (or dental decay). The stages of the dental decay are as follows:

*   Food (particularly food containing sugar) becomes lodged between the teeth.

*   Bacteria settle in the sugary deposits, using them for their own metabolism.

*   The bacteria excrete acids that dissolve the outer non-living covering (enamel) of the tooth.

*   A cavity develops, in which more sugary deposits collect – attracting more bacteria. More excreted acids increase the size of the cavity.

- Eventually, the decay reaches the living parts of the tooth – first the dentine, where there are nerve endings so the tooth begins to ache, then the pulp cavity – leading to an abscess.

Plaque

If the teeth are not regularly cleaned, a mixture of food, saliva, cheek cells and bacteria collects at the neck of the tooth (where it enters the gum). This deposit is called plaque and the acids released by the bacteria damage the enamel of the tooth in that region as well as cause gum disease.

Tartar (also called 'calculus')

If not removed, plaque hardens to form tartar. This forms a solid protection for the bacteria trapped between it and the tooth allowing them to decay the tooth without interference.

How to care for teeth

1 Do not eat sweet (or starchy) foods before going to bed as saliva flow stops when we are asleep preventing the sugar from being washed away. It is better to eat a raw crunchy food such as a carrot.

2 Brush the teeth last thing at night and first thing in the morning. Careful brushing removes plaque.

3 Use dental floss regularly to remove fragments of food from between the teeth.

4 Use a toothpaste that:

contains fluoride to strengthen the tooth enamel,

contains a bactericide to kill bacteria, and

is alkaline to neutralise acids released by the bacteria.

5 Visit the dentist regularly for an examination (and have treatment if required).

Progress check 7.4

1 What is the only non-living component of a tooth?

2 Which teeth are responsible for grinding food?

3 Why should a toothpaste be alkaline in nature?

7.04 A summary of chemical digestion in the alimentary canal

The need for chemical digestion

As most foods contain insoluble starch, protein and fat, it would be impossible for them to be absorbed into the blood stream if the large and insoluble molecules were not broken down into small, soluble and **absorbable (diffusible)** ones. Glucose, amino acids, fatty acids and glycerol – all represent the smallest forms of their 'parent' molecules (Table 7.3).

Name of food (carbohydrates)	Where digested	Enzyme involved	Product
starch	buccal cavity	amylase	simple sugars
starch	duodenum	amylase (from pancreas)	simple sugars
proteins	stomach	protease stomach walls also secrete hydrochloric acid that (i) kills bacteria and (ii) creates the optimum pH for the action of stomach protease	amino acids
proteins	duodenum	protease (from pancreas)	amino acids
proteins	duodenum	protease	amino acids
fats	duodenum	lipase (from pancreas and intestinal walls)	fatty acids and glycerol

Table 7.3 Chemical digestion of food

Worked example

a Distinguish between mechanical and chemical digestion

b Explain how a student who is standing on his head is still able to swallow food down into his stomach.

Answer

a This part is requiring you to realise that food is taken into the mouth in large pieces which have to be broken down into smaller pieces before the alimentary canal can process them. This is achieved in two ways: (i) using teeth, and (ii) churning the food in the stomach. These are both forms of mechanical digestion. The teeth chop (incisors and canines) the food, which is also ground up by the molars (and premolars). Proteins, such as in meat, need to be chopped. Cell walls in plant material need to be ground up to release cell contents.

The stomach then churns the food into a soup after which the original food has a maximum surface area exposed ready for the next stage, which is chemical digestion.

Chemical digestion is the process during which large insoluble chemical molecules are broken down into small, soluble chemicals using enzymes. Proteins become amino acids as a result of the action of proteases, fats become fatty acids and glycerol, due to lipase and carbohydrases, such as amylase, digest large carbohydrate molecules to simple sugars, such as glucose, which can be absorbed by the walls of the ileum.

b This part of the question is aiming to produce a description of peristalsis, but you need to explain that peristalsis is a muscular action and can therefore work against gravity. A bolus of food is swallowed into the oesophagus and the circular muscles of the oesophagus contract and close the oesophagus behind it. This contraction then passes in a wave-like motion, along the oesophagus towards the stomach, pushing the bolus in front of it. Once the bolus has passed, the circular muscle relaxes and, aided by contraction of the longitudinal muscles, the oesophagus is opened in preparation for receiving the next swallowed bolus.

Name of food (carbohydrates)	Where digested	Enzyme involved	Product
starch	buccal cavity	amylase	maltose
starch	duodenum	amylase (from pancreas)	maltose
maltose	on the membranes of the lining (epithelium) of the small intestine	maltase	glucose
proteins	stomach	a protease known as pepsin that works only when the pH is acidic (low) – caused by the HCl	polypeptides (incompletely broken down proteins)
proteins	duodenum	a protease called trypsin (from pancreas) that works in a higher pH (salts from the liver in bile and from the pancreas neutralise the HCl)*	polypeptides
polypeptides	small intestine	protease	amino acids
fats	duodenum	lipase (from pancreas and intestinal juice) in the presence of bile, made in the liver and released into the duodenum via the bile duct	fatty acids and glycerol

Table 7.4 Digestion of food in the alimentary canal

*The HCl also destroys harmful bacteria that may enter the alimentary canal in food by denaturing their enzymes.

A more detailed summary of chemical digestion in the alimentary canal

More information is given about the digestion of foods and the final products in Table 7.4.

The functions of bile are:

- to help to **neutralise acidic chyme** from the stomach and thus produce a suitable pH for the enzymes of the duodenum to operate

- to **emulsify fat** (i.e. break the fat into tiny droplets – increasing their surface area and therefore also the rate at which they are digested by lipase).

Progress check 7.5

1 Without reference to the book, write a list of the digestive enzymes, where each is produced in the digestive system and what job each enzyme does. Then check your list against Table 7.4.

2 The digestion of which food stuff in particular would be affected if a person had their gall bladder removed and why?

The part played by the villi in the absorption of food

The villi of the ileum are specially adapted for the process of food absorption. They:

- are extremely numerous – increasing the internal surface area of the ileum

- are very thin-walled (covered with a single layer of cells – the epithelium)

- have their cell membranes thrown into minute folds (**microvilli**) to further increase their surface area

- contain blood capillaries just beneath their walls

- contain special structures (lacteals) for absorbing fatty acids and glycerol

- are able to move to bring themselves into close contact with food.

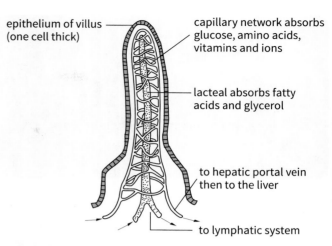

Figure 7.5 A villus

Amino acids and glucose after absorption by the blood capillaries are carried by a blood vessel, called the hepatic portal vein, directly to the liver for the first stage of their treatment in the body. Fats, after entering the lacteals, travel in the lymphatic system, by-passing the liver and enter the circulatory system at a vein in the neck.

Assimilation

The main food substances, absorbed as small soluble molecules, must now be built up into the larger molecules needed by the body.

Glucose (and any other simple sugar that can be absorbed by the villi) may be used, as it is as a substrate for respiration, to release energy. After a meal, there will be more glucose available than is needed immediately for this purpose. Therefore, it needs to be stored. Hence, it is built up into a large insoluble molecule called glycogen (a polysaccharide like starch). It is stored in the cells of the liver and muscles under the effect of the hormone **insulin** from the pancreas. When the blood sugar concentration falls below the required level, glycogen can then be reconverted to glucose (under the effect of hormones **adrenaline** and **glucagon**) and released into the blood.

Amino acids are used in cells for building up proteins as the cells grow and also for making special proteins such as enzymes. Blood proteins, in particular, are made in the liver. Amino acids (as well as proteins) are never stored. Any in excess to our needs at the time, are broken down in the liver by a process called **deamination**.

Chapter summary

◻ You have learnt the constituents of a healthy diet, can state a source for each one and know their functions in the body.

◻ You will now know the parts with their functions of the alimentary canal.

◻ Also, where amylase, protease and lipase are produced, where they are found and what their functions are in the alimentary canal.

◻ You have learnt the structural features of a tooth and how to care for your teeth.

Exam-style questions

1   Table 7.5, when complete, lists some of the food materials that are digested, the enzymes that digest them and the end products of their digestion.

a   Copy and complete Table 7.5.

food material	enzyme	end products
	amylase	
		amino acids
		fatty acids and glycerol

Table 7.5   [6]

b   Describe the processes that occur in the liver when there is:

i)   an excess of amino acids in the blood [3]

ii) a lack of glucose in the blood. [2]

[Total 11]

2   Figure 7.6 shows a villus from the small intestine, and a cell taken from the villus.

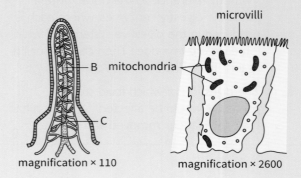

magnification × 110   magnification × 2600

Figure 7.6

a   i)   State the region of the villus from which the cell is taken. [1]

ii) State the function of Structure B. [2]

iii) Name the blood vessel into which the contents of Structure C are carried. [1]

b   Explain the importance of the mitochondria present in the cell taken from the villus. [4]

[Total 8]

3   Explain how food is caused to move along the alimentary canal. [10]

Chapter 8

Transport in plants

Learning outcomes

By the end of this chapter you should understand:

- The names, positions and functions of conducting tissues in plant organs

- How plants lose water and the factors affecting their water loss

- How to conduct experiments to demonstrate the effects of these factors

8.01 Transport in plants

The transport system of a flowering plant has to provide means of carrying water and ions from the roots to the leaves. This occurs in veins, known as **vascular bundles**.

Vascular bundles contain **two** transport tissues:

- **Xylem** for carrying water and ions, and

- **Phloem** for carrying sucrose and amino acids.

TIP

It will help you to remember what is carried in the xylem and phloem if you think of the three consecutive letters of the alphabet 'W, X and Y', '**W**' stands for **W**ater in the **XY**lem. Phloem carries the plant's food substances sugar and amino acids (**PH**loem carries the plant's **PH**ood)!

The movement of sugar and amino acids around a plant is called translocation.

The positions occupied by xylem and phloem in the vascular bundles of roots, stems and leaves in a typical dicotyledonous plant are shown in Figure 8.1.

Xylem is in the form of long tubes that stay open because their walls are strengthened with the chemical lignin. We are more familiar with lignin and its strength in the form of wood. Lignified xylem vessels therefore also:

- help support the stem and photosynthesising tissues of the leaf, and

- resist forces on the root that might cause the plant to be pulled out of the ground.

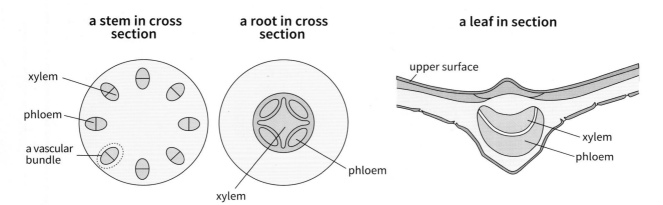

a stem in cross section   **a root in cross section**   **a leaf in section**

Figure 8.1 The position of xylem and phloem in a stem, root and leaf (to remember the position of the xylem, remember that the centre of a circle is often marked with X)

8.02 Water uptake

In the section on osmosis (Chapter 2), the methods by which water enters root hairs was explained, root hair structure was shown in Figure 2.9 and the functions stated:

- The absorption of water.
- The absorption of mineral ions in solution.
- The absorption of oxygen, in solution, for root respiration.

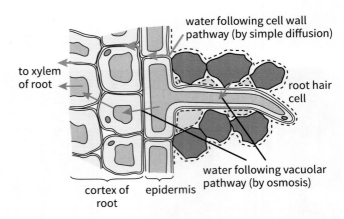

Figure 8.2 Water entering root hairs and crossing the cortex of a root

Water travels across the **cortex** of the root (the cells between the epidermis and the central vein) and enters the **xylem** in the central vein (the 'stele') in the root.

It then:

- travels up the xylem in the root, into the xylem first in the vascular bundles of the stem
- then into the xylem in the veins in the leaves
- and is then transferred to the **mesophyll cells** of the leaf.

Progress check 8.1

1   Which nutrient chemical is carried in the xylem?

    A   amino acid

    B   glucose

    C   nitrate

2   What is carried in the phloem?

Ensure that you are familiar with the section in Chapter 2 dealing with the appearance of root hairs and the ways in which root hair cells are adapted to their functions. Note, particularly, the section that refers to the uptake of ions by root hairs using the process of **active transport**.

PRACTICAL

This water movement can be demonstrated practically as follows:

    Aim: **To show that water travels up a stem in the xylem**

    Apparatus: A beaker

                  Food colouring

                  A soft-stemmed dicotyledonous plant (e.g. a broad bean or a stem of balsam. Celery works very well, but a 'stick' of celery is really a leaf stalk, not the true stem of the celery plant)

Method:

Cut the plant stem about 1 cm above the root and place it in water containing food colouring and leave, as shown, for about 1 day. Remove the stem from the beaker. Carefully, cut it through 2–3 cm from its base.

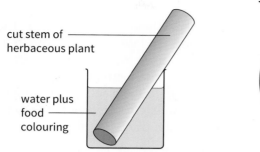
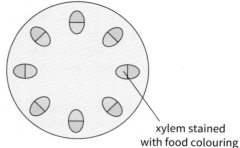

section through stem after 6 hours

cut stem of herbaceous plant

water plus food colouring

xylem stained with food colouring

Figure 8.3 Demonstration that water travels in the xylem of a stem

Results:

The vascular bundles will be seen to have been stained by the food colouring. If a very thin section is carefully cut from the stem, using a razor blade and viewed under a microscope, it will be seen that the coloured region of the vascular bundle is in the position occupied by the xylem.

8.03 Transpiration

Transpiration may be defined as **the loss of water vapour by evaporation of water at the surfaces of the mesophyll cells followed by the loss of water vapour through the stomata.**

Transpiration occurs as follows:

1 Water leaves the xylem vessels in the leaf where it forms a **water film on the walls of the mesophyll cells** inside the leaf. This water film is for dissolving carbon dioxide for use during photosynthesis.

2 **Water** from the water film **evaporates into the intercellular (leaf) spaces** – greatly increasing the humidity of the air inside the leaf.

3 There is now a greater concentration of **water vapour molecules** inside the leaf than in the atmosphere outside, so they **diffuse out through the stomata** into the atmosphere.

TIP

Transpiration does **not** occur from the leaf surface, but from the surfaces of the cells inside the leaf.

So long as the stomata are open for the uptake of carbon dioxide, then transpiration is bound to occur. If its rate exceeds that at which water can be absorbed from the soil, then water starts to be lost from the plant's cells. The cells lose their turgidity, the leaves and stem begin to droop and the plant is said to wilt.

Conditions affecting the rate of transpiration

Those atmospheric conditions that affect the rate of evaporation of water also similarly affect the rate of transpiration as shown in Table 8.1.

Transpiration speeded up by	Transpiration slowed down by
dry air	humid air
high temperature	low temperature
bright light*	dim light*

Table 8.1 Conditions affecting transpiration

*This condition does not affect the rate of evaporation, but affects the size of the stomata through which the water vapour passes.

Worked example

Two similar well-watered potted plants, one with the upper surfaces of its leaves and one with the lower surfaces of its leaves coated with Vaseline (petroleum jelly), are left in a warm sunny and dry place.

Explain what you think would happen to them over the following days if they did not receive any further water.

Answer

To start with, both plants would receive enough water to allow them to photosynthesise and to lose water by transpiration. The plant that has Vaseline on the upper surface of its leaves continues to transpire, taking water from the soil, and to photosynthesise, taking carbon dioxide from the air. It will continue to look healthy for several days until the soil dries out. The plant will then begin to lose more water more rapidly than it can replace it from the soil. Its cells lose the pressure inside them (their turgidity) causing a lack of support to the plant, which droops (the correct word is 'wilts'). In due course, this would lead to the death of the plant.

However, the plant whose leaves have Vaseline on the lower surface will not be able to transpire because its stomata, which are on the lower surface of a leaf, are blocked by petroleum jelly. It would not be able to absorb carbon dioxide through its stomata. When the other plant is wilted, this plant will still be firm and upright. The inability to photosynthesis would not be noticed for a good deal longer, when it would first cease growing, then, some time later, it would die through lack of nutrition.

PRACTICAL

The effects of temperature, humidity and light intensity on the rate of water uptake by a leafy twig can be investigated using apparatus called a (bubble) potometer, which can be made in the laboratory.

Apparatus: A length of capillary tube

2–3 cm of rubber tubing that fits tightly over the capillary tube

A syringe

A ruler to act as a scale

A beaker of water

A clock with a seconds hand

A stand and clamps for support

A clear polythene bag, large enough to fit comfortably over the shoot

Several bench lamps

3 bunsen burners

A thermometer

Material: A smooth-stemmed leafy shoot that will fit tightly into the hole in the rubber bung

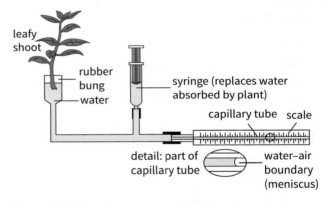

Figure 8.4 A potometer made in the laboratory

Method:

The experimental control is set up as in the following description. Set up the apparatus as shown in Figure 8.4. The capillary tube and the leafy shoot should be assembled under water in a sink to ensure that no air is trapped in the apparatus.

The apparatus is then placed on a laboratory bench and left for 15 minutes to settle, after which, the syringe is used to return the meniscus back to the end of the capillary tube. As water is absorbed by the shoot, the water/air boundary (the meniscus) in the capillary tube moves towards the shoot. The meniscus can always be returned to the end of the capillary tube by gently pushing more water in from the syringe.

After the shoot has acclimatised to the conditions, measure and record the length of time taken by the meniscus to travel a given distance as shown by the scale. Return the meniscus and repeat the procedure until you have at least three records of the time taken for the bubble to travel the same measured distance. Calculate the average of the three readings.

To investigate the effect on transpiration of each of the following conditions, the potometer should first be used as described previously (the control) – to give a set of results to compare with the experimental ones obtained by the following modifications:

1 To show the effect of humidity on the rate of water uptake:

Carefully enclose the leafy shoot in the polythene bag. Leave the shoot to acclimatise for 15 minutes, then take at least three further readings as before. Record each and calculate the average.

Results and conclusion:

The water vapour leaving the shoot will increase the humidity within the polythene bag. The time taken for the meniscus to travel the measured distance should be longer than the time taken with no polythene bag over the plant (the control). The increased humidity decreases the rate of transpiration. This in turn, decreases the rate of water uptake by the shoot.

2 To show the effect of light on the rate of water uptake:

With the polythene bag removed, direct the light from at least three bench lamps onto the shoot. Leave for 15 minutes to acclimatise and repeat the procedure of recording three readings and taking the average.

Results and conclusion:

The time taken for the meniscus to move the measured distance should be shorter than the control. In bright light, the stomata under the leaves will open further allowing water vapour to escape more quickly.

3 To show the effect of heat on the rate of water uptake:

This time, the shoot is moved to a warmer (or cooler) atmosphere. This may be achieved by using two different rooms – one with air conditioning/central heating and the other without. If it is not possible to arrange for a similar level of illumination, then, in both rooms, illumination should be by three, similarly placed bench lamps After a few minutes, record the temperature. If rooms of different temperature are not available, a temperature difference may be obtained with the use of several Bunsen burners. Leave the apparatus for the usual 15 minutes. Take three or more readings and record the temperature when each reading is taken. Calculate both the average time and the average temperature of the readings.

Results and conclusion:

The time taken should again be shorter than in the control. Water evaporates more quickly in higher temperatures, increasing the rate of transpiration, thus increasing the rate of water uptake.

1   The potometer can only measure the rate of water uptake. It cannot directly measure the rate of transpiration. Some of the water taken up will be used in photosynthesis in the leaves and not released during transpiration.

2   Some of these experiments can be conducted using potted plants in well-watered soil. The pots and soil are enclosed in a polythene bag to prevent evaporation and the pots are then weighed on scales before and after the experiments. The difference in mass is a direct indication of the amount of water lost by the plant.

Progress check 8.2

1   What extra information could be obtained using the calibrations on the syringe?

2   How are stomata related to the movement of water in a plant?

   A   water enters through them as water vapour

   B   water enters through them in liquid form

   C   water leaves through them as water vapour

   D   water leaves them in liquid form

The mechanism by which water is carried from the soil to the leaves

Osmosis is responsible for water entering the root hair cells and for its passage across the cortex cells to the xylem. It creates a force (root pressure) helping to push water up the xylem to the vascular bundles of the stem.

Once in the xylem of the stem, water is carried upwards by the second force at work – that of capillarity. This is the tendency for liquids to travel upwards through very narrow tubes (and can be demonstrated by dipping the end of very narrow capillary tube in water). Capillarity is the result of forces of molecular attraction between water molecules and molecules in the walls of the xylem vessels. Xylem vessels have a microscopic bore that can be responsible for carrying water 20 cm, or more, up the plant.

The third force results from the evaporation of water from the mesophyll cells of the leaf and the removal of that water through the stomata of the leaf (creating a water potential gradient). This creates the force known as transpiration pull as it draws up more water to replace that has been lost.

The three forces causing water to rise up a plant are therefore:

1   root pressure

2   capillarity

3   transpiration pull

Transpiration pull relies for its effectiveness on the fact that (water) molecules are attracted to one another and are thus not pulled apart as they move up the plant. These forces of 'molecular cohesion' ensure a continuous stream of water and ions travelling up the plant, known as the transpiration stream.

8.04 Translocation

Translocation may be defined as the movement of sucrose and amino acids in the phloem.

As a plant photosynthesises during daylight hours, it manufactures glucose, which, in many plants, is turned to starch and stored, temporarily, in the chloroplasts. In order to make more room for glucose and starch production, the stored starch in converted to a soluble sugar (sucrose) so that it can be carried away (translocated) to specialised storage organs such as stems or roots where it is usually converted to starch and stored, or to the growing regions of the root or shoot where it is converted to glucose for energy release during respiration. Carriage of the carbohydrates is always in the phloem.

Nitrogen (in the form of nitrates in solution in the xylem) is used in the leaves to manufacture amino acids, which are also translocated in the phloem, to the growing regions where they are used to manufacture proteins.

Translocation in the phloem is thus responsible for carrying sucrose and amino acids both up and down a plant depending on the plant's requirements at the time.

Variations in translocation

Where the stored or manufactured chemical is moved from is referred to as the **source** and where it moves to, and gets used, is known as the **sink**. Thus the same organ can be either a source or a sink depending on the time of day and the consequent direction of movement of chemicals.

During daylight hours, in a healthy, growing plant, the **source** of the carbohydrates and amino acids will usually be the leaves while the **sink** will be in the lower, non-photosynthesising regions of the plants (the roots). Carbohydrates and amino acids will be needed in the roots as follows:

Carbohydrates

- For release of the high levels of energy required during **growth** and **active transport**.
- For **storage** (often as starch) as an energy reserve.

Amino acids

- For building proteins to make cytoplasm in the new cells and often for some **storage**.

During the non-photosynthesising hours of darkness (or cold), the source and sink **become reversed**. The stored carbohydrates and proteins are now in greater concentration in the roots and are need for growth processes in the shoots. At such times, the phloem will carry sugars and amino acids upwards towards the leaves.

Progress check 8.3

1   What is the difference between transpiration and translocation?

2   Explain why the leaves of a plant can sometimes be a source and sometimes a sink, with respect to carbohydrates.

Chapter summary

- ☐ You now know the differences in the functions of xylem and phloem and how these tissues are arranged in a root and in a stem.

- ☐ You have also learnt where transpiration occurs and the effect of climatic factors on it.

- ☐ You have carried out experiments that show the effects of climatic factors on transpiration.

- ☐ Students following the supplementary course have learnt which chemicals move during translocation and the factors that control this movement.

Exam-style questions

1 Figure 8.5 is a transverse section through part of a young dicotyledonous plant.

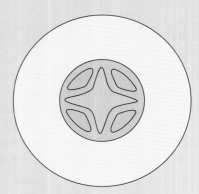

Figure 8.5

a State from which part of the plant the section has been cut. [1]

b On Figure 8.5, label and name: **i)** the xylem; **ii)** the phloem. [2]

c State the contents of the xylem and phloem and describe the pathway followed by these contents in order to be present in the positions shown in Figure 8.5.

xylem contents _____

_____

_____

pathway _____

_____

_____ [4]

phloem contents _____

_____

pathway _____

_____

_____ [3]

[Total 10]

2 a Describe what happens to the water that enters a leaf up to the point at which it enters the atmosphere. [8]

b Explain the factors that ensure that a steady stream of water is supplied to a leaf. [5]

[Total 13]

3 The graph shows how the rate of water uptake by a plant changes under different environmental conditions.

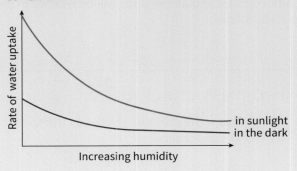

Figure 8.6

Using information in the graph:

a Explain the change in the rate of water uptake in a plant in sunlight as the surrounding atmosphere gradually becomes more humid, [3]

b Explain the differences in the rate of water uptake under similar atmospheric conditions, when the same plant is placed in darkness. [4]

[Total 7]

Transport in animals

9.01 Circulatory system

When applied to animals, the term 'transport system' refers to the **circulatory system**. As in the flowering plant, it comprises a system of tubes (blood vessels) but, unlike the flowering plant, this time there is a **pump** and a system of **valves** to ensure that the liquid within the tubes (blood) flows through them in only one direction.

The single circulation of a fish

Since a very important function of the blood is to carry oxygen, there must be some stage in the circulation during which oxygen is taken from an animal's environment into its blood. This occurs at what is known as the animal's **respiratory surface**. In the case of fish, this is provided by the gills. Blood is pumped by the heart to the gills, picks up oxygen, then travels under relatively low pressure distributing oxygen to all the internal organs before returning to the heart containing carbon dioxide. The heart pumps the blood under pressure to the gills where it loses its carbon dioxide and picks up a fresh supply of oxygen. In one circulation of the body, the blood therefore passes once through the heart, and the fish is said to have a **single circulation** (Figure 9.1).

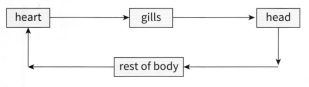

Figure 9.1 Single circulation of a fish

Double circulation

During one complete circulation of the human (mammalian) body, blood travels twice through the heart – once on its way to the lungs having arrived in the heart from the other organs of the body and, the second time, on its way to the other organs having arrived from the lungs (Figure 9.2).

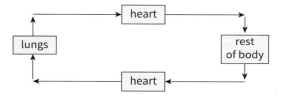

Figure 9.2 Double circulation of a mammal

The heart is completely divided into two halves in order to play its part in the dual circulation. The right side of the heart deals with **deoxygenated** blood and the left side deals with **oxygenated** blood. Since the lungs lie either side of the heart, the distance the blood has to travel is short compared with how far it may travel in order to reach other parts of the body. Blood leaving the heart in the lesser circulation (to the lungs) is not under so much pressure as it is in the greater circulation. (Blood must always be able to reach the tissues of the finger tips, even when the arms are raised.)

A double circulation ensures that fully oxygenated blood is sent quickly, under high pressure, to all the organs of the body. It is also quickly returned to the heart and then to the lungs to lose its carbon dioxide and become fully oxygenated again. There is no mixing of oxygenated with deoxygenated blood.

1 In a single circulation, during which part of the circulation is the blood under the most pressure?

2 What is meant by the term **double circulation**?

A Blood passes through the heart twice in one complete circulation.

B Blood travels twice round the body before being pumped to the lungs.

C The heart contains two different types of valve.

D There are two types of chambers in the heart, atria and ventricles.

9.02 The heart

Important facts about the heart are given here.

1 It is a muscular pump.

2 It has four chambers, all with similar volumes when full – two **atria** 'on top' (in mammals that walk upright) of two ventricles.

3 Atria (singular atrium) receive the blood in **veins** (venae cavae are large veins).

4 **Ventricles** have thick muscular walls to pump the blood out of the heart into **arteries** under pressure.

5 A system of valves ensures one-way flow of blood through the heart (see Figure 9.3a).

6 As the ventricles contract, the mitral (or bicuspid) and tricuspid valves are slammed shut causing the 'lubb' sound of the heartbeat.

7 The heart muscle then relaxes and as the ventricles do so, the semi-lunar valves close to stop blood being drawn back into the ventricles, causing the 'dupp' sound. There is then a pause before the next cycle (and the next 'lubb–dupp' of the heart).

8 The left and right sides of the heart are separated by a wall called the septum. The **septum** ensures that oxygenated and deoxygenated blood do not mix.

Unlike the ventricle walls, the atrial walls have only a small amount of muscle and their walls are thin as they have only to send blood to the ventricles below. The left ventricle has the thickest walls for sending blood round the body in the aorta (the body's largest artery). The right ventricle has only to send blood to the nearby lungs.

When the heart beats, a wave of contraction passes through the heart from the atria to the ventricles. It is triggered by electrical impulses from a structure in the wall of the right atrium made of muscle and nerve endings and called the pacemaker. The atria contract just before the ventricles.

During one heart beat (or 'cycle'), the following events occur:

- The walls of the atria contract.

- Blood passes into the ventricles through the tricuspid and bicuspid valves that are forced open by the pressure of the blood.

- The ventricles contract.

- Blood pressure closes the tricuspid and bicuspid valves and opens the semi-lunar valves.

- Deoxygenated blood is forced out of the right ventricle and travels to the lungs via the pulmonary artery and, at the same time, oxygenated blood is forced out of the left ventricle and is sent off round the body in the aorta.

- The atria relax, drawing a fresh supply of deoxygenated blood from the venae cavae into the right atrium and oxygenated from the pulmonary vein into the left atrium.

- The ventricles relax ready to receive blood from the atria.

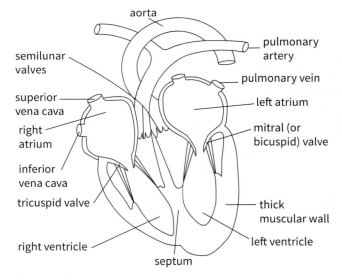

Figure 9.3a The mammalian heart – ventral view

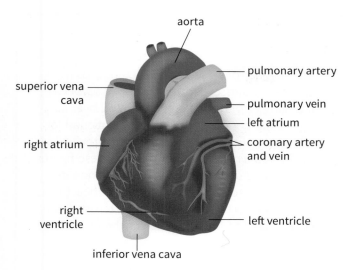

aorta

pulmonary artery

superior vena cava

pulmonary vein

left atrium

coronary artery and vein

right atrium

right ventricle

left ventricle

inferior vena cava

Figure 9.3b External view of heart

The names of the one-way valves need to be known only for those following the Syllabus Supplement.

Progress check 9.2

1   Name the blood vessel carrying blood from the lungs to the heart

2   What is directly responsible for the sound made by the heart in one heart beat?

   A   the mitral and tricuspid valves closing just before the semilunar valves

   B   the mitral valve closing just before the tricuspid valve

   C   the semilunar valves closing just before the mitral and tricuspid valves

   D   the tricuspid valve closing just before the mitral valve

3   Sit quietly and still, with your right leg crossed over your left leg. You will notice that your foot 'kicks' very slightly and regularly. Can you explain why this is?

Monitoring the activity of the heart

For a healthy life, it is important to ensure that you have a healthy heart. There are various ways of checking whether the heart is beating normally:

* By feeling the pulse. This gives an idea how strongly the heart is beating and how quickly it is beating (the average rate in an adult human being around 70 beats per minute).

* By listening to the sound of the heart. Its normal 'lubb-dupp' sound may be accompanied by other sounds ('heart murmers') if the valves are not closing properly, allowing blood to leak back into the atria or ventricles.

* By taking an **ECG**. This is a recording of the electrical impulses that control the contraction of the heart muscles and can detect abnormalities that may need to be monitored closely or treated (with drugs or surgery).

Each beat of the heart creates a surge of pressure in the arteries (the pulse). There is no such surge in the veins where blood flows smoothly under much lower pressure. However, veins have semi-lunar valves, which ensure that blood continues to flow back towards the heart but, otherwise, veins rely on the movement of nearby muscles (e.g. the calf muscles of the leg) to 'massage' the blood from one set of semi-lunar valves to the next.

Counting the pulse

The pulse beat in an artery can be located by gently pressing on the wrist at the base of the thumb using your middle and index fingers (Figure 9.4).

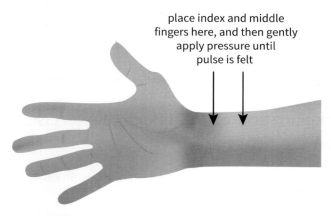

place index and middle fingers here, and then gently apply pressure until pulse is felt

Figure 9.4 Counting the pulse

TIP

This skill may take a little practice. Don't press too hard, as this stops blood passing along the (radial) artery.

The number of pulse beats in 1 minute can be counted but it is more usual to count the number of beats in a shorter period of time, such as 15 seconds, and then calculate the rate per minute.

PRACTICAL

Aim: **To study the effect of physical activity on pulse rate**

Method:

The number of pulse beats is counted over three separate minutes. Each result is recorded and an average taken.

About 5 minutes exercise is performed (e.g. running or stepping on and off a stair step).

The number of pulse beats in 10 seconds is counted and recorded, then, after 20 seconds, the number of beats is again counted over a 10-second period. The process is continued for 10 minutes. Each recorded number is multiplied by 6 to obtain the rate per minute. These figures can then be used to draw a graph of the rate against time to show how the rate of heart beat gradually returns to normal after exercise (as in Figure 9.5). The fitter the subject, the quicker the pulse rate returns to normal.

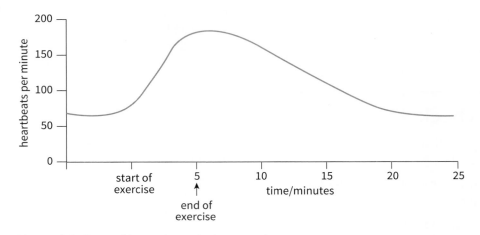

Figure 9.5 Rate of heart beat during exercise

Explanation of the effect of exercise on the heart

Use of muscles during exercise increases the amount of carbon dioxide in the blood.

Higher levels of carbon dioxide in the blood are detected by that part of the brain that controls heart rate; it responds by stimulating the pacemaker situated in the heart muscle to increase the rate at which it sends out impulses to the muscle that surrounds it. The heart thus beats faster.

Other factors also increase the heart rate, including a higher pressure of blood reaching the heart and an increase in the hormone adrenaline in the blood stream.

The faster heart rate ensures:

- a faster removal of the extra carbon dioxide from the blood travelling to the lungs
- a faster rate of uptake of oxygen to supply the increased activity in the muscles.
- a more rapid supply of glucose to muscle cells for greater energy release during increased respiration.

Coronary heart disease

People who lead stressful lives are also at risk of heart disease, since stress causes the release of raised levels of the hormone adrenaline, which has the effect of constricting artery walls that may already be partially blocked – further increasing the blood pressure.

To decrease the risk of heart disease, therefore:

1   Restrict the intake of animal products containing fat as animal fats may be deposited on artery walls – particularly those of the coronary artery, restricting blood flow. Also avoid too much salt (NaCl) as this tends to increase blood pressure.

2   Avoid obesity – extra weight puts extra strain on the heart.

3   Do not smoke – nicotine encourages the depositing of fat on the artery walls.

4   Settle for a less stressful life-style.

5   Take regular exercise.

There is a particular need to take these measures if there is a history of heart disease in your family, as deficiencies affecting the heart are often inherited. Heart attacks occur more commonly in people over the age of 45 and, according to some research, men are over 30% more likely to suffer a heart attack than women.

Diet, exercise and the treatment of heart disease

A large proportion of the fats found associated with meat and meat products are of a type known as **saturated** fats. Also present in meat is the fatty substance **cholesterol**. With time, saturated fats and cholesterol build up on the walls of our arteries (forming **atheroma**) and cause occlusion of the blood vessel (block the flow of blood). If this happens in the coronary artery, it prevents oxygen reaching parts of the heart muscle causing a 'heart attack'.

Since the heart is a muscle, then, like all other muscles in our bodies, in order to remain healthy, it should

be regularly exercised. Activities that raise the heart rate for at least a few minutes daily are thus strongly recommended in order to maintain a healthy heart.

Progress check 9.3

1   Figure 9.6 shows a section through the heart (viewed from the front). Which valve closes in order to ensure that blood flows to the lungs?

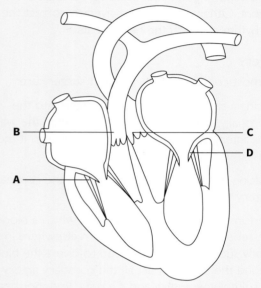

Figure 9.6 Rate of heart beat during exercise

2   Which valve closes to prevent oxygenated blood being drawn into the heart?

3   List the causes of heart disease.

Worked example

a   What is it that is by-passed in heart bypass surgery and why?

b   Veins are often used to form the bypass. Why must care be taken over which way round the inserted vein is joined to the damaged blood vessel?

Answer

a   It is the coronary artery that is by-passed because it has become blocked by deposits on its wall. The deposits contain animal fats.

(Supplementary candidates should also say that the deposit is called atheroma and also contains cholesterol.)

b   Veins have valves that allow blood to flow in one direction only – back to the heart. In this operation, the vein would be taking blood from the heart and thus the vein would have to be the correct way round to ensure that the pressure of the blood does not close the valves in it.

The treatment of heart disease

Depending on the particular heart problem experienced by a patient, treatment may involve drugs and/or surgery.

Drugs

Drugs (heparin, warfarin and aspirin) may be used to help to prevent or break down blood **clots** blocking the coronary artery. 'ACE inhibitors' help to widen blood vessels supplying the heart, and to decrease blood pressure. Other drugs can be given to correct the rate and rhythm at which the heart beats.

Surgery

The most common forms of heart surgery are:

* Using a small balloon that is inserted into the coronary artery, then inflated to stretch the walls of the artery allowing blood to flow through it more easily.

* Placing of a wire cage (a **stent**) into the coronary artery to hold it open.

* Performing a 'bypass' operation in which a blood vessel – usually a vein taken from elsewhere in the body such as the leg – is used to divert the blood round the blocked part of the coronary artery and, if absolutely necessary and if a suitable donor heart is available, performing a heart transplant.

9.03 Blood and lymph vessels

Blood always leaves the heart in **arteries** that are vessels with thick muscular walls, able to withstand the high pressure of the blood. Blood returning to the heart under much lower pressure does so in thinner-walled vessels called **veins**. The largest artery is called the aorta; a small artery is called an arteriole. The largest vein is the vena cava; a small vein is a venule. Blood passes from arterioles to venules through microscopic blood vessels called **capillaries**.

Figure 9.7 shows the names of the blood vessels associated with some of the main organs of the body.

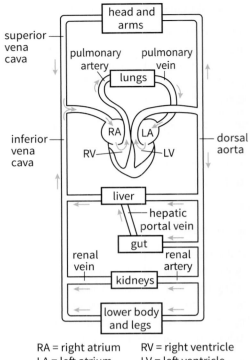

RA = right atrium   RV = right ventricle
LA = left atrium   LV = left ventricle

Figure 9.7 The mammalian circulatory system

A comparison of the three types of blood vessel

Blood travels round the body as shown in Figure 9.8.

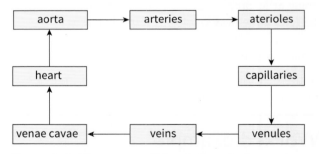

Figure 9.8 The route of blood through the body

The structure and functions of arteries, veins and capillaries

Blood circulation is maintained by:

- the force of contraction of the ventricles of the heart

- the muscles and elastic fibres in the arteries that prevent the arteries from ballooning as the pulse beat passes (a 'recoil' effect)

- valves in veins that prevent backflow

- the contraction of skeletal muscles near veins, which help to push the blood from one set of valves to the next

- the 'suction' (reduced pressure) caused each time the heart relaxes.

Arteries

Arteries are required to take oxygenated blood with blood sugar and other materials to all the organs and, thus, have to maintain blood pressure so the blood reaches all organs no matter how far away from the heart they may be. The thick muscular and elastic walls help to maintain this pressure.

Veins

Veins carry blood that has lost its pressure after passing through capillaries. They rely on their one-way, semilunar valves and on their thin walls that allow the effect of muscular contractions to be passed to blood within, pushing it from one set of valves to the next.

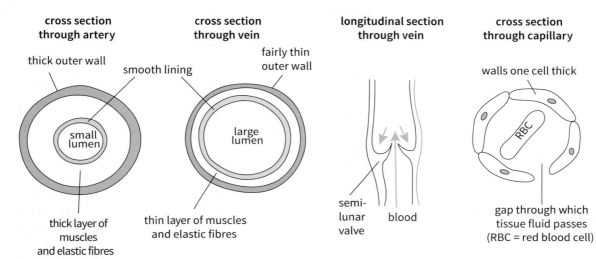

Figure 9.9 A comparison of blood vessels

Arteries	Veins	Capillaries
take blood from the heart	take blood to the heart	take blood from arteries to veins
blood under high pressure	blood under low pressure	pressure rises then gradually falls
blood flows in pulses	blood flows smoothly – no pulse	pulse gradually disappears
thick muscular walls with narrow lumen*	thinner walls with wide lumen	wall one cell thick, leaky and red blood cells travel in single file
no semi-lunar valves (except the pulmonary artery and the aorta)	semi-lunar vales present	no semi-lunar valves
carry oxygenated blood (except the pulmonary artery)	carry deoxygenated blood (except the pulmonary vein)	blood loses its oxygen (or gains it in the lungs)

Table 9.1 Differences between arteries, veins and capillaries

*The lumen is the hole through which the blood passes along the blood vessel.

Capillaries and the transfer of materials between them and tissue fluid

Capillaries allow blood plasma together with its dissolved materials, blood **platelets** and white blood cells (but not red blood cells and blood proteins) to escape through their thin 'leaky' walls (Figure 9.9). This fluid is called **tissue fluid**, which bathes all body cells, supplying them with oxygen, glucose and so on and maintains them at the optimum concentration (due to osmosis). The network of capillaries is so dense that all cells are situated close to a capillary. Carbon dioxide (from cell respiration) passes into the tissue fluid, some of which returns to the capillaries, the rest being collected in vein-like vessels called **lymphatic vessels**, which take the fluid, now called **lymph**, to rejoin the circulation via a vein in the neck (collecting on the way, lymph containing fat from the lacteals of the villi).

A particular network of capillaries can be by-passed in emergencies (e.g. to parts of the skin exposed to cold conditions) by the use of **shunt vessels** that connect arteries directly to veins.

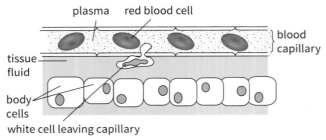

Figure 9.10 The relationship between blood, tissue fluid and body

Lymph nodes

Situated at intervals along the system of lymphatic vessels are swellings called **lymph nodes** (or **lymph glands**). They contain an accumulation of the white blood cells called **lymphocytes** (see the section on blood next). Lymph nodes also **filter bacteria** from the blood, thus the lymph system plays an important part in the body's defence against infection (the **immune system**).

During a medical examination, the lymph glands are often felt to see if they are enlarged, which may be an indication of an infection or some other condition affecting the immune system.

9.04 Blood

Structural features and functions of the individual components of blood are discussed here.

Blood is made up of four basic components:

1 Red blood cells
2 White blood cells
3 Platelets
4 Plasma

Red blood cells

These are by far the most numerous of the cells in the blood. They are disc-shaped, but thinner in the middle than they are round the edges. Because there are so many of them and because they are so very small, this considerably enlarges their surface area for absorbing **oxygen**, which they then transport around the body. They contain the orange/red pigment **hemoglobin**, which is able to absorb oxygen and carry it as **oxyhemoglobin**.

White blood cells

These cells help to defend the body against infections by microorganisms. Some produce proteins called **antibodies** that are specifically designed to stick to the outer walls of microorganisms preventing them from absorbing nutrients and removing their excretory products. In this weakened state, the microorganisms are then **ingested** by other white blood cells by a process called **phagocytosis**.

Platelets

These are fragments of cells that plug any gaps caused by damage to capillary walls and are responsible for triggering the process of blood clotting.

Plasma

This is the pale yellowish fluid in which the blood cells and platelets are carried. Its job is to transport substances round the body:

- excretory products – carbon dioxide from cells to the lungs, urea from the liver to the kidneys

- food substances – amino acids, glucose, vitamins and ions after they have been absorbed from the alimentary canal to all cells of the body.

The plasma:

- contains proteins such as the antibodies from white blood cells and fibrinogen for blood clotting

- carries hormones
- carries heat.

Details of the blood cells are shown in detail in Figure 9.11.

A prepared slide, viewed under a light microscope, is shown in detail in Figure 9.12.

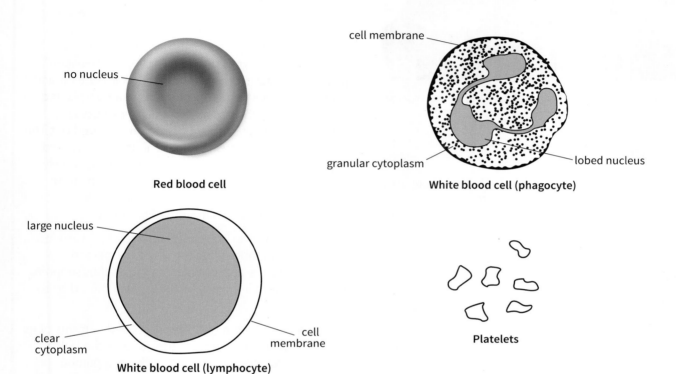

no nucleus

Red blood cell

cell membrane

granular cytoplasm

lobed nucleus

White blood cell (phagocyte)

large nucleus

clear cytoplasm

cell membrane

White blood cell (lymphocyte)

Platelets

Figure 9.11 Blood

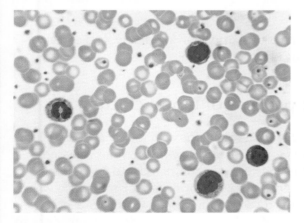

Figure 9.12 Blood as seen under a microscope

Progress check 9.4

1   Can you identify the types of blood cell in Figure 9.12 and explain why do most of the cells appear to have holes in the centre?

2   Why haven't the proteins found in blood plasma been digested to amino acids?

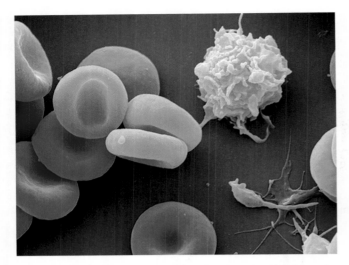

Figure 9.13 Photomicrograph of red and white blood cells and platelets

Red blood cells

They are unusual cells in that they have no nucleus. As a result, they have a limited life span (120 days).

White blood cells

There are two types:

1 Phagocytes
2 Lymphocytes

Phagocyte structure

They have a lobed nucleus and are capable of movement ('amoeboid' movement) and can squeeze out from capillaries.

Function of phagocyte

They carry out phagocytosis – that is, they move towards potentially harmful bacteria that have been immobilised by antibodies and surround them (the correct terms are 'engulf' or 'ingest' them), then digest them, thus preventing or overcoming infection.

Lymphocyte structure

They have a large round nucleus occupying almost the entire cell.

Functions of lymphocyte

They produce antibodies, which are proteins that 'stick' to bacteria and clump them together ready for being ingested by **phagocytes**. Some antibodies are in the form of **antitoxins**, which neutralise poisons (toxins) in the blood that have been released by invading bacteria.

Antibody production and tissue rejection

Antibodies are specific to the organism against which they are produced. They are proteins that may stay in the blood only for a few weeks or for a lifetime, giving life-long immunity against the effects of that particular disease-causing agent (or pathogen). However, since our immune system is unable to distinguish between possibly harmful proteins, as in pathogenic bacteria, and potentially useful 'foreign' proteins, as in a transplanted heart or kidney, there is always a danger of **tissue rejection** when a transplant operation is carried out. The more similar the protein structure in the transplanted organ is to the proteins of the recipient, the less likely are the chances of rejection. Organs, therefore, from a (close) relative are far less likely to be rejected, since there will be a greater similarity of protein type.

Vaccination involves subjecting a person to a harmless strain of the pathogen, against which the person manufactures the appropriate antibodies, thus supplying the person with immunity to the disease caused by the pathogen.

The lymphatic system has been described in terms of its returning tissue fluid and fats absorbed by the lacteals to the circulatory system. It also includes a number of lymph glands that are responsible for the manufacture of lymphocytes. These glands can become swollen and painful as they manufacture lymphocytes when a person is fighting an infection.

Pus that collects around an infected wound contains many dead white blood cells.

The process of blood clotting

Fibrinogen is a soluble protein found in blood; when it comes in contact with enzymes released by damaged cells, it is converted into an insoluble, stringy protein called **fibrin**. This forms a mesh which traps blood cells and becomes a clot – preventing the entry of pathogens. The clot dries and hardens to form a **scab**, which covers the wound until the skin beneath has repaired.

Chapter summary

■ You have learnt the structure of the heart.

■ You have learnt about the structure of the different types of blood vessel.

■ You have learnt about the structure and functions of the components of blood.

■ Students following the supplementary course have learnt about single and double circulation.

■ Students following the supplementary course have learnt how the heart functions.

Exam-style questions

1 a State how

   i) a person's life-style [5]

   ii) and other factors, contribute to heart disease. [3]

2 Figure 9.14 is a diagram showing a heart-lung machine in use.

  a On Fig. 9.14, use label lines to identify:

   i) the left atrium

   ii) the tricuspid valve

   iii) the right ventricle. [3]

b Name the blood vessel:

   i) from which the heart-lung machine takes blood

   ii) into which it returns the blood. [2]

c Explain the use of the term 'heart-lung' as applied to this machine. [5]

[Total 10]

3 Explain how the structure of the circulatory system keeps blood moving continuously round the body. [5]

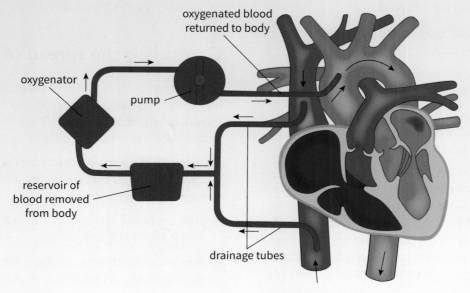

oxygenated blood returned to body

oxygenator

pump

reservoir of blood removed from body

drainage tubes

Figure 9.14

Diseases and immunity

10.01 Disease

Disease can have a variety of causes, but a common cause of disease is the invasion of body tissues by another, usually microscopic organism known as a **pathogen**.

A **pathogen** can thus be defined as a **disease-causing organism**.

A common feature of pathogens is that they can be passed from one person ('host') to another. **A disease caused by a pathogen that can be passed from one host to another** is known as a **transmissible disease**. Transmissible diseases can be sub-divided into:

- those that pass during direct body contact with an infected person (e.g. HIV/AIDS)

- those that pass indirectly:

  - through the air (e.g. influenza)

  - from surfaces or food touched by other people (e.g. cholera)

  - from animals (e.g. rabies and various forms of alimentary infection).

If the disease is caused by a pathogen, such as a bacterium or a virus, then the pathogen has to overcome a succession of natural defences that the body possesses.

1 First, the skin is a **mechanical**, bacteria-proof covering. Fresh or open wounds should therefore always be protected until the natural barrier is repaired by first a blood clot and then a scab. Cuts should be carefully washed with a disinfectant to kill any bacteria that have been able to gain entry. Air passing down into the lungs has first to pass through the nose, where hair filters out potentially harmful pathogens.

2 The body produces **chemicals** to trap and kill pathogens before they have chance to establish themselves in the body tissues. Mucus in the respiratory tract (nose, bronchi and bronchioles that form the passage to the lungs) is a solution of protein to which pathogens become stuck before they reach the more delicate tissues of the lungs. Pathogens swallowed in food and drink almost immediately have to survive a very low pH caused by hydrochloric acid in the stomach, which kills the vast majority of them. Tears are mildly antiseptic.

3 As we have seen in the earlier section on blood, any bacteria that survive either or both of defences 1 and 2, then have to survive the effects of **antibodies** produced by some of the **white blood cells**, and being engulfed and digested (**phagocytosis**) by others. Vaccination, when the body is exposed to a relatively harmless form of the pathogen, causes the white blood cells to manufacture antibodies that are then ready to act on the relevant pathogen if it invades the body at a later time.

Controlling the spread of disease

To stay clear of disease is impossible, but there are a number of measures that can be taken both by the individual as well as by all members of the community to decrease the spread of disease.

As many diseases are contracted through infected foods, it is essential that proper hygienic practices surround food preparation.

- Surfaces on which food is prepared should be washed with antibacterial solutions.

- Raw meat, which often contains high levels of pathogens, should never come into contact with cooked and cooled foods, or with foods that are consumed uncooked.

- Cooked foods should be kept in refrigerators (below 5 °C) to discourage bacterial growth.

- Those handling food should **always** wash their hands after visiting the lavatory and always before handling food.

- Organic waste should always be properly disposed of and, as faeces are a major source of pathogens, sewage should always be properly treated.

Progress check 10.1

1   List the ways by which pathogens can pass from one person to another.

2   State how the body is protected against bacterial invasion.

10.02 Immunity

How antibodies work to kill pathogens

The presence of pathogens within the body causes lymphocytes in the blood to manufacture chemicals called **antibodies** (Figure 10.1). Antibodies have '**variable ends**' that are 'tailor-made' to fit the molecular shape of the particular pathogen. The antibodies thus 'lock' onto the pathogen. This serves three purposes

1   It stops the pathogen entering the cells of the body.

2   It prevents any further development of the pathogen leading to its destruction.

3   It clumps the pathogens together ready for destruction by phagocytes.

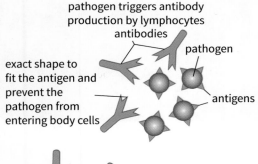

pathogen triggers antibody production by lymphocytes

antibodies

pathogen

exact shape to fit the antigen and prevent the pathogen from entering body cells

antigens

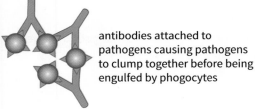

antibodies attached to pathogens causing pathogens to clump together before being engulfed by phogocytes

Figure 10.1 How antibodies attack pathogens

Active immunity

The production by the body of antibodies to protect against invading pathogens is called active immunity – defined as **defence against a pathogen by antibody production in the body**.

The body can produce its own antibodies either in response to a pathogen or in response to a vaccination (when the body is exposed to a relatively harmless form of the pathogen). In both cases, it is the chemical molecules on the surface of the pathogen that are responsible for the exact form of antibody being produced. The molecules on the pathogen's surface are called **antigens** (see Figure 10.1). Once antibodies have been produced, the body is said to have gained **immunity** from the pathogen. The ability to produce appropriate antibodies is retained by lymphocytes called **memory cells** and, thus, immunity from the same pathogen is retained for life.

Antibodies produced against some pathogens have short-lived effectiveness, as the pathogen is always rapidly mutating and is thus no longer destroyed by the antibody produced when it last infected the person (e.g. those produced against the influenza virus).

The **effectiveness of vaccination programmes** has been demonstrated by the fact that many dangerous diseases can now controlled (e.g. poliomyelitis) and some, notably smallpox, have been wiped out completely.

Passive immunity

This is defined as **resistance to a pathogen through antibodies that are passed from one individual to another, such as from mother to child**.

In the case of passive immunity, it is not the infected person who manufactures the antibodies, thus no memory cells are involved and consequently, immunity is not long-lasting. Passive immunity is particularly important for an infant who is being fed on its mother's milk, as in this way, it is protected against infection until its own immune system becomes fully active. Indeed, protection began while it was still in the uterus as antibodies pass from mother to fetus through the placenta along with dissolved food and oxygen.

Antibodies made by some other organisms may be introduced to provide passive immunity, but immunity lasts only a short time. For example, antitoxins from horses are used against bacterial food poisoning (botulism) and were often injected into humans to counteract the bacteria that cause tetanus.

When the immune system does not serve us well

It is unfortunate that, sometimes, the antibodies produced attack the person's own otherwise healthy cells. Such a condition is called an **autoimmune response**. This can be a cause of type 1 (or juvenile) diabetes when, sometimes triggered by the presence of a virus, the lymphocytes manufacture antibodies that destroy the cells in the pancreas (in the islets of Langerhans) that make insulin.

TIP

1   Never say that antibodies or phagocytes 'kill' diseases. They act against the pathogens that cause disease.

2   Never say that pathogens become immune to antibodies. It is the person who makes the antibodies that becomes immune to the pathogens.

Worked example

Explain the reasons for vaccinations, such as those against measles, lasting a lifetime, while others, such as the vaccination against influenza, need to be given every year in order to be effective.

Answer

This question is really asking two things: (i) how vaccinations work, and (ii) why do they not work so well against some illnesses.

First, both are examples of active immunity. That is immunity that is a result of a person being exposed to a weakened form of the pathogen that causes the disease (a virus, in both cases) and they actively manufacture the antibodies that are tailor-made to render the virus inactive. Lymphocytes, known as memory cells, retain the ability to manufacture the same antibodies very rapidly if they meet the same pathogen in the future. Some pathogens, particularly those that are less common, are always likely to be affected by the antibodies but some pathogens, often the more common ones such as the influenza virus, are always undergoing mutations and thus, each year, it is a slightly different strain that spreads throughout the population. The person retains immunity against the previous year's strain, but the antibodies are ineffective against the mutated strain, thus a fresh vaccination has to be given and this vaccine will have been manufactured using the new, mutated virus. There is also the problem that many strains of the influenza virus may be circulating at the same time. A person may have some resistance to one strain but not to another.

Chapter summary

■ You have learnt how infectious diseases are spread.

■ You have also learnt about the body's natural defences, including active immunity and how antibodies work.

■ You have learnt about the difference between active and passive immunity.

Exam-style questions

1 a   State the measures that can be taken to prevent the spread of pathogens. [6]

   b   i)   Explain what is meant by *immunity*. [3]

       ii) Explain how immunity is achieved. [5]

2 Explain how white blood cells protect a person from a bacterial disease. [8]

Respiration and gas exchange

11.01 Gas exchange

It is essential for the respiration of cells of the body that they are constantly supplied with oxygen from the atmosphere and the waste carbon dioxide is rapidly removed and expelled into the atmosphere. This exchange of gases between the body and the atmosphere takes place in all organisms at gas exchange surfaces, which include gills (fish), the skin (amphibians) and the lungs.

Mammals, such as human beings, have lungs containing many minute air sacs called **alveoli** (Figure 11.1).

Alveoli are adapted for the process of gaseous exchange in the following ways:

1 The millions of alveoli provide a large surface area for gaseous exchange.

2 The walls are only one cell thick for quick and easy diffusion of gases in solution.

3 The walls have a layer of moisture to dissolve gases.

4 They are richly supplied with capillaries for rapid transport of the gases.

5 There is an efficient pumping mechanism to bring fresh supplies of oxygen and remove the carbon dioxide.

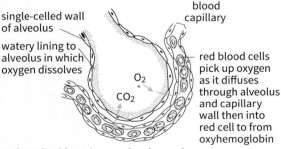

single-celled wall of alveolus

watery lining to alveolus in which oxygen dissolves

blood capillary

red blood cells pick up oxygen as it diffuses through alveolus and capillary wall then into red cell to from oxyhemoglobin

O_2

CO_2

carbon dioxide arrives at alveolus as the hydrogencarbonate ion in the plasma and then diffuses into the alveolus to be expelled during exhalation (breathing out)

Figure 11.1 Gaseous exchange at alveolar surface

The structure of the respiratory organs

A pair of lungs lies in the (airtight) thoracic (chest) cavity, as shown in Figure 11.2.

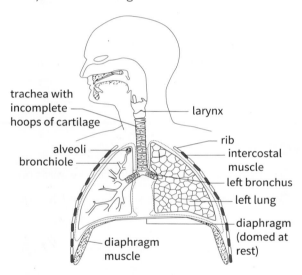

trachea with incomplete hoops of cartilage

larynx

alveoli

bronchiole

rib

intercostal muscle

left bronchus

left lung

diaphragm (domed at rest)

diaphragm muscle

Figure 11.2 The respiratory organs (only candidates following the supplementary course are required to know the following labels: incomplete hoops of cartilage, internal and external intercostal muscles, diaphragm)

Oxygen enters the blood from the lungs. In order to move air containing oxygen from the atmosphere to the exchange surfaces of the lungs, a muscular, pumping action called breathing must occur. Breathing in (inhalation or **inspiration**) is responsible for presenting air with its oxygen to the surfaces where gaseous exchange will take place. Exhalation (or **expiration**) pushes the air, now containing waste carbon dioxide, back into the atmosphere (breathing out).

The differences between inspired and expired air

Inspired and expired air differ in several ways, as shown in Table 11.1.

Inspired air	Expired air
20% oxygen	16% carbon dioxide
0.04% carbon dioxide	4% carbon dioxide
relatively dry	saturated
at air temperature	at body temperature
relatively dirty	relatively clean

Table 11.1 Comparison of inspired and expired air

Progress check 11.1

1 Why do you think that the thoracic cavity must be airtight?

2 What happens to a red blood cell as it passes in a capillary through the lungs?

During gas exchange in the lungs, approximately the same amount of oxygen is taken into the blood as the amount of carbon dioxide that leaves it. This follows from the equation for respiration that shows the same number of molecules of oxygen being used up as of carbon dioxide released by the process.

The lining of the alveoli, the bronchioles and the bronchi are moist. The moisture in the alveoli is used for dissolving the oxygen. The air in the bronchioles and bronchi is moist due to the solution of mucus lining the tubes. The heat of the body ensures that water is evaporating into the exhaled air. Hairs in the nose and mucus in the respiratory tract collect dust particles, causing the air breathed out to be cleaner than that breathed in.

The control of rate and depth of breathing

The link between rate and depth of breathing and the pH of blood

Respiration in the cells produces carbon dioxide. Carbon dioxide, in solution, forms carbonic acid.

Worked example

a **Distinguish between breathing and respiration.**

b **Describe how alveoli are suited to their function.**

Answer

a This is a fairly common question since many incorrectly believe that respiration is the technical term for breathing. Two clear descriptions are required. Breathing is a muscular action that takes air into the lungs for the uptake of oxygen and expels air from the lungs removing carbon dioxide. Respiration is a chemical reaction in cells during which glucose is oxidised to release energy. There are two forms of respiration: aerobic, which takes place in the presence of oxygen and releases relatively large amounts of energy and anaerobic, which takes place in the absence of oxygen and releases comparatively small amounts of energy.

b It is necessary first to state the function of the alveoli. This is the process of gaseous exchange, specifically the uptake of oxygen and its passage from the air in the lungs to the blood in the capillaries and the removal of carbon dioxide from the blood to the air in the lungs. Then you should consider how the structural features of the alveolus aid the process of gaseous exchange. First, oxygen in the air in the lungs must dissolve and, for this, the lining of the alveolus is moist. The oxygen has then to travel into the blood. This requires passing through the alveolus wall, which is only one cell thick to reduce the distance the oxygen has to travel. The alveolus is in close contact with many capillaries – again to reduce the distance that the oxygen has to diffuse. The features of the alveolus described for oxygen passage apply also to carbon dioxide diffusing in the opposite direction. Finally, there are millions of alveoli, each surrounded by a network of capillaries providing an extremely large surface area for the maximum uptake of oxygen and release of carbon dioxide.

- The carbonic acid (as the **HCO$_3^-$** ion) passes into the blood, **lowering its pH**.

- **Receptors** (chemoreceptors) in the **brain** (**medulla**) and in the **aorta**, sensitive to the blood's pH, control the rate and depth of breathing.

- During **exercise, more carbon dioxide** is released. This causes the blood's pH to fall – a change that is detected by the receptors, which initiate nervous impulses that are passed to the **intercostal** and **diaphragm muscles** to increase their rate and degree of contraction.

As the additional carbon dioxide is expired, the pH of the blood gradually returns to its resting level and so does the rate and depth of breathing.

PRACTICAL

Many features of respiration can be demonstrated practically.

Aim: **To show that air breathed out contains more carbon dioxide than air breathed in**

The presence of carbon dioxide can be shown either by using hydrogencarbonate indicator or limewater (Figure 11.3). Limewater turns from **clear** to **cloudy** in the presence of carbon dioxide.

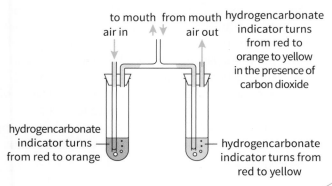

person breathing gently

to mouth air in / from mouth air out

hydrogencarbonate indicator turns from red to orange to yellow in the presence of carbon dioxide

hydrogencarbonate indicator turns from red to orange

hydrogencarbonate indicator turns from red to yellow

Figure 11.3 Experiment to show that air breathed out contains more carbon dioxide than air breathed in

Depending on the need for oxygen in the body (and the need to remove waste carbon dioxide), the rate and depth of breathing may vary.

More oxygen is used and more carbon dioxide is released when we are active. This can be demonstrated as follows.

Aim: **To show the effect of exercise on the rate of breathing**

Count the number of breaths taken in 1 minute by a person at rest. Remember, it is easier to count someone else's breathing rate than it is to count one's own.

Do this three times and take an average (around 15 breaths per minute). Perform 5 minutes, exercise as when investigating pulse rate. Using the same procedure as then, count the number of breaths in a 10-second period, every 30 seconds for 10 minutes.

Calculate the rate per minute and plot a graph of breathing rate against time.

Aim: **To show the effect of exercise on the depth of breathing**

The taking of regular exercise can affect the volume of air a person is able to inspire then expire in one deep breath. This can be demonstrated by using the simple **respirometer** shown in Figure 11.4.

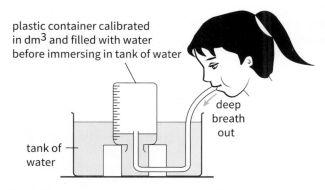

plastic container calibrated in dm^3 and filled with water before immersing in tank of water

tank of water

deep breath out

Figure 11.4 Measurement of the effective lung volume

A group of athletes and a group of non-athletes can, in turn, breathe in as far as they are able then breathe out through the rubber tube as far as they can. In each case, measure the volume of water expelled from the container. An average per person for the athletes and for the non-athletes can be calculated and the results compared. Athletes will be likely to be able to exhale a greater volume of air (and thus, be able to breathe more deeply).

The immediate effect of exercise on the depth of breathing may be successfully demonstrated only with a commercial respirometer, where the person breathes in and out through a tube connected to a piece of electronic equipment that provides a graph of the depth (as well as the frequency) of breathing. A graph similar to Figure 11.5 would be expected if a person measured the depth and rate of breathing before and every 30 seconds after a period of exercise.

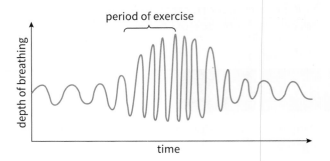

Figure 11.5 The effect of exercise on breathing rate and depth

11.02 How the lungs are ventilated (that is, how we breathe)

During breathing, the lungs are inflated and deflated by the action of muscles that help to form the thorax. They are:

- the internal and external intercostal muscles (between the ribs), and

- the muscles of the diaphragm (a domed sheet lying below the lungs, separating them from the abdomen).

These muscles work to alter the volume of the thoracic cavity, which, in turn, alters the pressure within the thoracic cavity.

During inspiration

1. The external intercostal muscles contract. This causes the ribs to swing up and out – increasing the volume of the thorax ('front' to 'back').

2. The muscles of the diaphragm contract, pulling it flat – increasing further the volume of the thorax (this time 'top' to 'bottom').

The resultant increase in volume of the thorax decreases its pressure.

Boyle's law states that for a fixed amount of gas at a fixed temperature, pressure and volume are inversely proportional – that is, when one doubles, the other halves.

The only communication with the higher pressure of the outside atmosphere is via the mouth/nose, so air from the atmosphere is forced towards the region of lower pressure, passing through the mouth/nose and down the trachea. The trachea must not be allowed to collapse and is held open by (incomplete) **hoops of cartilage**. They are incomplete so as to form a groove in which the oesophagus lies. Air passes on through the bronchi and bronchioles and into the alveoli. Gaseous exchange occurs.

During expiration

1. The external intercostal muscles relax – the ribs swing down and in.

2. The diaphragm muscles relax and the diaphragm domes upwards again.

These actions decrease the volume of the thoracic cavity, increasing its pressure. Air is forced back out into the atmosphere.

There are also (smaller) internal intercostal muscles, which relax during inspiration and contract during expiration to help provide some extra force to the action of breathing out.

How air is cleaned before entering the lungs

The trachea is **lined** with **ciliated cells** (see Chapter 1) and goblet cells. Goblet cells secrete mucus, a sticky fluid containing protein that traps bacteria and dust particles as they pass in inhaled (breathed in) air. The mucus is then carried towards the throat by the **upward-beating** microscopic projections (**cilia**) of the ciliated cells.

> **TIP**
>
> It may help if you think of the mucus as a 'moving carpet'.

The mucus, plus dust particles and bacteria are then swallowed. Many of the bacteria will perish when the meet the stomach acid. In this way, the lungs are protected against many harmful (pathogenic) bacteria.

> **TIP**
>
> Remember that the cilia do not act as filters for dust and bacteria – they keep the carpet of mucus moving.

11.03 Respiration

Respiration is a characteristic of all living organisms. It is defined as the **chemical reactions that break down nutrient molecules in living cells to release energy**.

The chemical reactions are reactions of metabolism, and are thus controlled by their own particular set of enzymes

> **TIP**
>
> Energy is always 'released', 'liberated', 'evolved' and sometimes 'converted'. It is never 'produced' or 'made', except by the Sun.

The human body uses this energy for the following processes:

1. for muscle contraction (bringing about movement and locomotion)

2. to help link together amino acids in order to manufacture proteins (such as enzymes and those used for growth and repair)

3. for cell division

4. for active transport of chemicals through cell membranes (e.g. the uptake of glucose through the villus walls)

5. conversion into electrical energy in the form of impulses along nerve cells

6. conversion into heat energy to maintain a constant body temperature.

There are two forms of respiration:

1. aerobic respiration

2. anaerobic respiration.

Aerobic respiration

Aerobic respiration is the **chemical reactions in cells that use oxygen to break down nutrient molecules to release energy**.

Aerobic respiration usually takes the form of the oxidation of glucose in the cytoplasm of living cells, controlled by enzymes. It unlocks the chemical energy in the glucose molecule, releasing it for metabolic activities, and releasing also the waste products carbon dioxide and water.

glucose + oxygen →
        carbon dioxide + water + energy released

$$C_6H_{12}O_6 + 6O_2 \rightarrow 6CO_2 + 6H_2O + \text{energy released}$$

PRACTICAL

Practical demonstrations

1 **To demonstrate that oxygen is taken up during respiration.**

The apparatus is set up as shown in Figure 11.6. Either invertebrate animals or germinating peas can be used in the experimental test-tube.

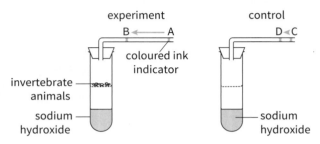

Figure 11.6 Apparatus to show uptake of oxygen in respiration

The experiment

For each molecule of oxygen taken in by the animals, a molecule of carbon dioxide is given out. The molecule dissolves in the sodium hydroxide, decreasing the volume within the apparatus. The indicator thus moves from A to B.

NB When all the oxygen in the tube has been used up, the total volume inside the apparatus would have decreased by 20%

The control

In the same period of time, the only carbon dioxide to dissolve in the sodium hydroxide is that already in the air. The indicator moves only from C to D.

2 **To show the effect of temperature on the rate of respiration**

The same apparatus is used, but this time the time taken for the indicator to move a set measured distance is recorded at three different temperatures.

First the test-tubes are places in an ice/water bath and left for 15 minutes before the measurement are taken. This is repeated twice more, the results recorded and an average calculated.

Then the same procedure is carried out in water baths, first at 20 °C, then at 40 °C.

Expected results:

The rate of movement of the indicator should be greater the higher the temperature.

Conclusion:

Oxygen is being used up faster as the temperature increases, the enzymes of respiration work faster with increased temperature, thus resulting in an increase in the rate of respiration.

Anaerobic respiration

Anaerobic respiration is defined as the **chemical reactions in cells that break down nutrient molecules to release energy without using oxygen.**

It is the process by which **yeast** cells break down glucose during **fermentation** to produce **ethanol** (alcohol).

glucose → alcohol (ethanol) + carbon dioxide + some* energy released

The balanced equation for respiration in yeast is:

$C_6H_{12}O_6 \rightarrow 2C_2H_5OH + 2CO_2 + $ some* energy released

Always check whether you are being asked about aerobic or anaerobic respiration.

*Less energy is released, per glucose molecule, in anaerobic respiration than in aerobic respiration since the alcohol molecule is relatively large and still contains a considerable amount of chemical energy.

Anaerobic respiration also occurs in the human body especially in **muscles**. It happens during vigorous exercise when there is not enough oxygen reaching the cells to convert all the glucose into carbon dioxide and water. Instead, the glucose is partially broken down, without oxygen, to lactic acid, with only a limited amount of energy being released.

glucose → lactic acid + some energy released

In muscles, lactic acid builds up if it is made faster than it can be carried away by the circulatory system. Excess lactic acid in muscle cells causes them to react by suddenly contracting during a painful bout of **cramp**. The lactic acid has to be oxidised to carbon dioxide

(i.e. it has to undergo **aerobic** respiration) – a process that happens in the in the liver to where some of it has been taken by the blood stream.

Since more oxygen is required by both muscle and liver cells than when normally at rest, the breathing rate and the heart rate remain increased for some time after the person has stopped exercising. Eventually the rates return to normal after the '**oxygen debt**' has been paid off.

Progress check 11.2

1. Why, do you think, gentle exercise is sometimes described as 'aerobic' exercise?

2. What is released in aerobic respiration?

    A alcohol

    B glucose

    C lactic acid

    D water

Chapter summary

■ You have learnt the structure of the lungs, including that of the alveoli.

■ You have learnt the difference between breathing and respiration.

■ You learnt how breathing takes place and the differences between inspired and expired air.

■ You have learnt the essential features in aerobic and anaerobic respiration.

Exam-style questions

1. Figure 11.7 shows the rate and depth of breathing of a person over a 20 minute period.

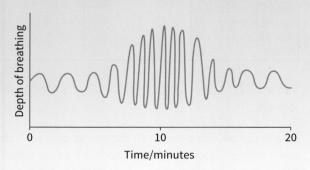

Figure 11.7

a. Using information from the graph, account for the changes shown in Figure 11.7 during the 20-minute period. [7]

b. i) State where the gas exchange surfaces are found in a human. [2]

   ii) List the features of human gas exchange surfaces that contribute to their efficiency. [3]

2. Describe and explain the events that occur in a muscle during a period of prolonged exercise. [6]

3. Explain what occurs in the thorax when a person breathes out. [7]

Excretion

12.01 Excretion in humans

The metabolic reactions within the body are always creating waste products. **Excretion** is defined as **the removal from organisms of toxic materials and substances in excess of requirements.**

There are two main sets of excretory organs for removing metabolic waste products; the lungs that excrete carbon dioxide and the kidneys that excrete **urea** and excess salts and water (Table 12.1).

Excretory organ	Material removed	Method of removal
kidney	urea	dissolved in water in urine
	excess salts	dissolved in water in urine
	excess water	depending on the amount of water drunk and the amount lost is sweat
lungs	carbon dioxide	breathing out (exhalation)

Table 12.1 Removal of waste by the kidneys and lungs

But note the following:

1. The sweat glands in the skin also remove a little urea since it is one of the constituents of sweat.

2. The removal of faeces from the alimentary canal is not regarded as excretion, since the fibre which makes up the faeces is not the product of a chemical reaction in the body. Instead, it is passed all the way through the intestines unaffected by any enzymes released. (Though the pigment which colours the faeces, which comes from the chemical breakdown of red blood cells, can be regarded as excretory.)

Kidneys

The kidneys are two brown bean-shaped organs lying dorsally (i.e. towards the back) in the abdominal cavity (Figure 12.1). They are supplied with blood through the renal artery and return it to the rest of the circulation through the renal vein. Their job is to filter urea and salts from the blood.

In order that the urea and salts can leave the body in solution, some water is also removed from the blood. The amount of water removed will depend on the factors shown in Table 12.2.

Factor	Effect on urine volume	Effect on urine concentration
larger volume of water consumed	larger volume of urine produced	urine more dilute
higher atmospheric temperature – more water lost in sweat	smaller volume of urine produced	urine more concentrated
exercise taken – body temperature rises – more sweat, less blood to the kidneys – less filtration of blood	smaller volume of urine produced	urine more concentrated

Table 12.2 Factors affecting removal of water by the kidneys

The materials removed from the blood by each kidney are then sent down the ureter to the bladder where they are stored. Relaxation of the bladder sphincter muscle allows them to leave the body as a solution called urine via the urethra.

TIP

urine = urea + ions + toxins + water

Make sure you learn the difference between urea and urine. Urea is a major chemical constituent of urine. Urine is the liquid containing urea that is removed from the body.

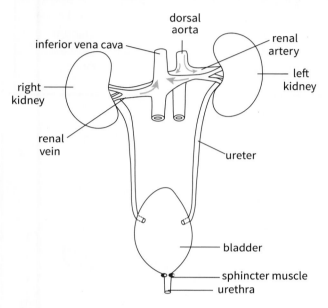

Figure 12.1 The excretory organs

The part played by the liver in the production of materials to be excreted

Urea

A balanced diet will always contain proteins whose constituent amino acids will be used by the cells to manufacture the particular proteins we need. There will then always be some **amino acids** that are in **excess** to our requirements. These are not stored, but **changed by the liver** to **urea**.

Progress check 12.1

1   Name the blood vessel that links the kidney with the (inferior) vena cava.

2   Why is the removal of faeces not considered to be an important form of excretion?

The fact that the liver deals with the amino acids that have been absorbed from the alimentary canal, means that it is therefore involved in their **assimilation**. During their assimilation, the required amino acids are linked together to make proteins, some of which will be **plasma proteins** that are then released into the blood.

The liver also deals with those amino acids that are not required to make proteins and are called 'excess' amino acids. During a process called **deamination**, it breaks them down into **urea**, which is released into the blood, and into a carbohydrate, which is converted into **glycogen** and stored in the cells of the liver.

Deamination is defined as **the removal of the nitrogen-containing part of amino acids to form urea**.

Urea, if allowed to build up in the blood, becomes toxic (poisonous), thus there must be a way of removing it. The same applies to carbon dioxide in the blood. Urea is removed by the kidneys while carbon dioxide is removed by the lungs.

Carbon dioxide has a narcotic effect on living cells. Fruit can be stored, after picking, in an atmosphere enriched with carbon dioxide to slow down the cells' metabolism and thus slow down the ripening process.

Worked example

a   **Explain what excess amino acids are and what happens to them in the body.**

b   **Explain why sweating and egestion of faeces are not normally regarded as excretory processes.**

Answer

a   This question covers three different topics – the chemicals of life, the alimentary canal and excretion. Amino acids are the basic units from which protein molecules are constructed. When we eat protein – such as meat – we ingest a range of different amino acids, some of which our bodies need in large quantities, some in only small amounts. If we take a lot of the amino acids that we require only in small amounts, then there is no point in keeping them – they are regarded as being 'in excess'. Some parts of the amino acid molecule can be converted into useful

→

carbohydrate and stored (as glycogen in the liver and muscles). The part that contains nitrogen is turned into urea in the liver by a process called deamination. The urea is carried in the blood to the kidneys where it is filtered out and removed from the body in urine.

b The important consideration here is to think of what is being removed from the body and why. In sweat, it is water that is being lost in order to control body temperature. The water is supplied to the sweat glands in the blood, which always contains some urea. Its loss in sweat is simply a coincidence, not a necessity. The removal of faeces is necessary since cellulose from the cell walls of plants is not digested by the human alimentary canal (though it does help to create bulk against which the muscles of the intestines can push, thus keeping the food moving along). It must therefore be removed, together, again coincidentally, with anything else the faeces contain. There is a case to be made for substances tipped into the alimentary canal as a way of removing them from the body (e.g. the remains of broken down red blood cells) being excreted, but be careful to understand the difference between excretion and excreta, which is a non-scientific term for faeces and urine.

The structure and functioning of the kidney

Each kidney is supplied with blood from the (dorsal) aorta along the **renal artery** and blood is returned from the kidney to the inferior vena cava along the **renal vein**.

The outer part of the kidney is the **cortex** and its inner part is the **medulla**. In the centre of the kidney is a space (the pelvis of the kidney) from which the **ureter** leads to the bladder.

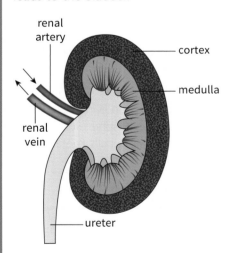

Figure 12.2 Longitudinal section of a kidney

In each kidney, the renal artery divides first into arterioles, then into capillaries. These capillaries form around a million minute clumps or 'knots'. The correct name for one of these knots is a **glomerulus**. Each glomerulus supplies a **kidney tubule** (or nephron) and each tubule starts with a cup-shaped structure called a **renal capsule** (Figure 12.3).

All renal capsules lie in the cortex of the kidney.

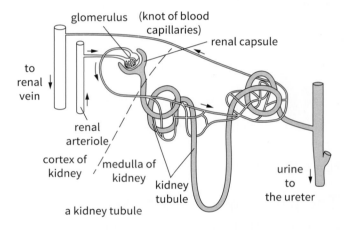

Figure 12.3 A kidney tubule

The knot of blood capillaries (glomerulus) in each renal capsule contains **blood** that, under the effect of the renal artery, is **under high pressure**. Under this pressure in the renal capsule, the blood in the knot of capillaries is forced out through the capillary walls and through the walls of the capsule into the renal tubule. This **high-pressure filtration** of the blood causes **water, glucose, urea, salts** and **toxins** to leave the blood and pass into the tubule.

Leading from the knot of capillaries, a capillary passes to the rest of the tubule and wraps itself in close contact around the tubule. As the filtrate passes through the tubule, the capillary absorbs all substances that the body cannot afford to lose. The following are reabsorbed into the blood (by a process called **selective reabsorption**):

• all of the glucose (unless the person is diabetic)

- most of the water (thus the urine in the tubule gradually becomes more concentrated)

- some of the salts.

This leaves in the tubule a concentrated solution of **urea**, **salts** and **toxins** otherwise known as **urine**.

NB The amount of water and quantity of salts reabsorbed can be adjusted depending of the amount consumed in the diet and/or the amount lost during sweating. This is **osmoregulation** ensuring that the blood remains at a more or less constant concentration.

Kidney dialysis

If the kidneys fail, then there will be a build-up of urea and toxins in the blood that would eventually prove fatal. Subject to availability and suitability of tissue type, a kidney transplant may be possible. Otherwise, **kidney dialysis** may be used. A kidney dialysis machine removes, from a patient's blood, chemicals with small molecules (urea, toxins and ions) but does not allow larger molecules (such as plasma proteins) or blood cells to leave the blood.

This is achieved by passing the patient's blood through **dialysis tubing** that is **partially permeable** (i.e. permeable only to the small molecules). This tubing is placed in a dialysis or 'washing' fluid, which is continuously renewed and washes away the substances removed from the blood (Figure 12.4). By varying the concentration of glucose and salts in the washing fluid, the amounts of those substances that **diffuse** from the blood can be controlled and, in this way, so can the concentration of the plasma in the blood that is then returned to the patient.

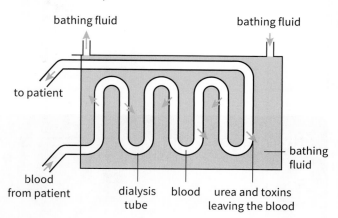

Figure 12.4 How an 'artificial kidney' works

Advantages and disadvantages of kidney transplants and dialysis

Kidney transplants and dialysis are compared in Table 12.3.

	Kidney transplants	Kidney dialysis
Advantages	1 It is a relatively permanent solution. 2 Patient can lead a comparatively normal life. 3 A less expensive procedure than dialysis. 4 Fewer restrictions on diet.	1 Dialysis provides an immediate solution to what could be a fatal condition.
Disadvantages	1 There is a shortage of suitable donors. 2 A transplanted kidney is made of foreign protein causing the production of antibodies that may lead to rejection of the organ. The transplant is more likely to be successful if the donor is a close relative with a similar protein type. 3 Anti-rejection drugs will always have to be taken.	1 The patient needs to undergo dialysis every 3 days or so. This is inconvenient and takes several hours each time. 2 There is always a risk of infection.

Table 12.3 Advantages and disadvantages of kidney transplants and dialysis

Progress check 12.2

1 In which structural part of the kidney do the glomeruli lie?

2 Which substance is filtered from the blood by the kidneys but does not appear in the urine of a healthy person?

   A amino acids

   B glucose

   C salts

   D urea

Chapter summary

■ You have learnt that the kidneys and the lungs are the body's main excretory organs.

■ You have learnt the structure of the kidney and the parts played by the kidney's structural features in the process of excretion.

■ You have also learnt how a patient with kidney failure can be treated.

Exam-style questions

1 Explain the factors that affect the volume and the concentration of the urine produced by a healthy person. [8]

2 The concentration of urea in a person's blood was measured before and during the time they began a course of dialysis treatment. The results are shown in Figure 12.5.

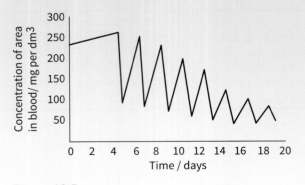

Figure 12.5

a Explain the blood urea concentrations before treatment. [2]

b i) State how long the first session of dialysis lasted. [1]

  ii) Calculate the decrease in urea concentration in the blood after the first session of treatment. [2]

c Explain why a kidney patient might be advised to restrict protein intake in their diet. [3]

d Explain how the loss of blood glucose during dialysis is prevented. [4]

[Total 12]

3 Why is excretion necessary and how is it carried out in the human body? [10]

Coordination and response

13.01 Nervous control in humans

The various organs of the body must be carefully **regulated** and work in **coordination** if an organism is to survive effectively in its environment. To achieve this, the body has a series of **receptors** that pass information about the environment to a coordinating centre, the central nervous system or **CNS** (i.e. the brain and the spinal cord). The CNS then directs a response in the appropriate **effectors** (**muscles** or **glands**) (Figure 13.1). The CNS is connected to the receptors and effectors by the peripheral nervous system.

Both the central and the peripheral nervous systems contain nerve cells (**neurones**).

> **TIP**
> Do not refer to 'nerves' when you mean 'neurones'. Nerves are bundles of neurones, like an electric cable contains separate wires. Neurones are individual nerve cells.

Neurones in the peripheral nervous system are **sensory neurones** and **motor** (or effector) neurones. The **sensory neurones** bring information from the receptors to the CNS (Figure 13.2). The motor neurones take information from the CNS to the effectors (Figure 13.3). The CNS contains **relay** (or connector) neurones linking the sensory and motor **neurones**. Motor and sensory neurones are elongated cells, with most of their cytoplasm concentrated in one place (the **cell body**), which contains the nucleus. The ends of a neurone are branched (the 'twigs' of the branches being called dendrites). The elongated part of a neurone (the **nerve fibre**) is insulated by a fatty sheath, and the fibre carries information along them in the form of (**electrical**) **impulses**.

> **TIP**
> Never say 'messages' when you mean 'impulses'.

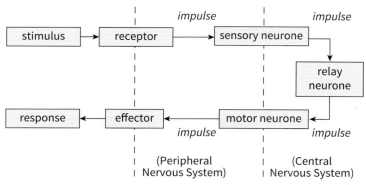

Figure 13.1 How a stimulus and a response are linked

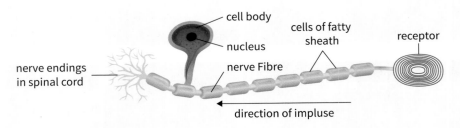

Figure 13.2 A sensory neurone

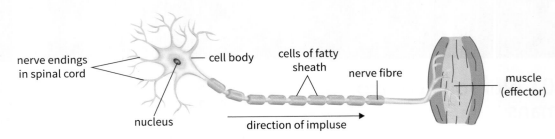

Figure 13.3 A motor neurone

Voluntary and involuntary actions

Voluntary actions

Voluntary actions are the result of a deliberate decision to move a part of the body. The decision is taken in the **brain** and **impulses** are sent along the **spinal cord** along neurones that link via **synapses** to the **motor neurone** that carries those impulses to the appropriate muscle. The muscle **contracts** and the desired **response** is achieved.

Involuntary actions

Involuntary actions result in automatic, fast and usually protective responses. **Reflex actions** are involuntary actions as they are **not controlled** by the brain. However, the brain is often aware that a reflex action is occurring. In reflex actions, the pathway followed by the impulses through the neurones is called a **reflex arc**.

Reflex actions are:

- rapid
- do not have to be learnt (i.e. are automatic)
- often, but not always, protective
- short-lived, and
- the same stimulus always results in the same response.

Reflex actions

There are occasions, often when danger threatens, when a sudden change in the environment (known as a **stimulus**) sets up, rapidly and automatically, a response or a set of responses in one or more effectors in the organism. These responses are coordinated with the stimuli so that the effectors (which are either **muscles** or **glands**) produce a response that is appropriate to the stimulus (Figure 13.4).

An example of a reflex action is the rapid removal of the hand if it accidentally touches a hot object and it occurs as follows:

- The hot object is the **stimulus**.
- The **receptor** is a temperature receptor in the skin containing modified nerve endings.
- An **impulse** is generated in a **sensory neurone**.
- The impulse arrives at the **spinal cord** (**CNS**).
- The impulses passes, in chemical form, across a small gap (**synapse**) between the dendrites (nerve endings) of the sensory neurone and the dendrites of a **relay neurone**.
- The relay neurone carries the impulse to a **motor neurone**.

- The motor neurone carries the impulse to the **effector** (biceps or triceps muscle in the arm, depending on the type of movement required).

- The muscle **responds** by **contracting** to lift the hand clear of the hot object.

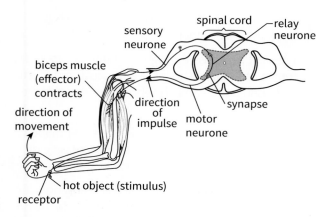

Figure 13.4 Diagram of a reflex arc

Progress check 13.1

1. i) Name the two types of effector.

   ii) State the difference between a nerve and a neurone.

2. What is the correct term for a sudden change in the environment?

Synapses

Although impulses appear to pass from neurone to neurone, the nerve endings of successive neurones do not touch one another. There is a small gap between them called a synapse. The impulse does not 'jump' the gap, but causes the neurone in which it has arrived to release a chemical called a neurotransmitter, which diffuses across the gap and stimulates receptor molecules in the membrane of the next neurone to generate another impulse. An example of a neurotransmitter is acetylcholine.

The definition of a **synapse** is a **junction between two nerve cells, consisting of a minute gap across which impulses pass by diffusion of a neurotransmitter**.

The neurotransmitter is contained within small bodies called vesicles, which, when the impulse arrives, are stimulated to release the neurotransmitter into the synapse, as shown in Figure 13.5.

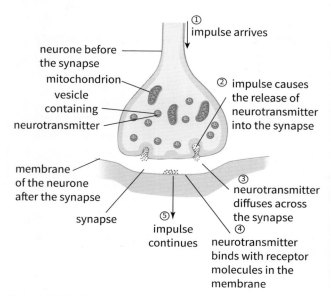

Figure 13.5 Transmission across a synapse

Since it is nerve endings of the neurone that bring the impulse that release the neurotransmitter, the impulse can pass in **one direction** only across the synapse.

The effect of drugs such as morphine and heroin on synapses

Morphine and heroin are chemicals with a molecular shape similar to neurotransmitters and thus they bind to the same receptor molecules at synapses as the neurotransmitters. This reduces the ability of a neurotransmitter to generate an impulse in the receiving neurone. This has a calming effect and is likely to be responsible for the feeling of euphoria experienced by takers of morphine and heroin. Its suppression of impulses also explains why morphine is used medicinally as a pain killer.

Progress check 13.2

1. Explain why impulses pass in only one direction across a synapse.

2. List the differences between an involuntary action and a voluntary action.

Worked example

Explain why there is a concentration of mitochondria in the nerve endings of a sensory neurone where it meets a relay neurone.

Answer

Mitochondria are associated with the process of respiration and, therefore, with the release of energy and energy is required to fuel the activities that occur at a synapse. First, the neurotransmitter substance has to be manufactured and stored in vesicles. The vesicles have to be moved to the sensory neurone's membrane, then opened to allow the neurotransmitter to be released into the synapse. This represents more activity than would be found in most body cells and therefore there is a concentration of mitochondria in this area.

13.02 Sense organs

Sense organs are defined as **groups of receptor cells responding to specific stimuli: light, sound, touch, temperature and chemicals**.

The sense organs that respond to sound waves are found within the inner ear. Those responding to touch and temperature are found in the dermis of the skin. The response to chemicals is a property of the cells of the taste buds in the tongue. Those responding to light are found in the retina of the eye.

The eye

The eye is one of the most important of the receptors providing us with information on dimensions, colours and distance away of objects in our environment.

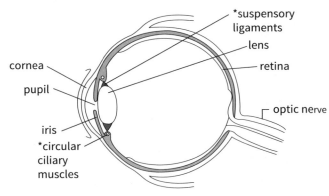

Figure 13.6 The eye.

*These labels are required only in the syllabus supplement.

How the eye functions to produce a focused image

1 Light rays from an object enter the transparent cornea.

2 The cornea 'bends' (**refracts**) the light rays in towards one another.

3 The light rays pass through the **pupil** (a gap in the centre of the iris. The iris is a muscular structure that controls how much light passes through the pupil)

4 The transparent, elastic **lens** is altered in shape. It is made either fatter or thinner, decreasing or increasing (respectively) its focal length.

5 The relatively small amount of refraction now produced by the lens brings the rays to **focus** on the retina.

6 The **retina** contains receptors in the form of light sensitive cells, some of which are sensitive to light of **different colours**. These cells are stimulated by the light of the image, and convert the light energy into electrical energy. The most sensitive part of the retina on which objects in the centre of the field of vision are focused is called the **yellow spot** or fovea.

7 Electrical energy, in the form of an **impulse** travels along the **optic nerve** to the **brain**.

8 The brain decodes the impulses to produce the sensation of sight.

Rod and cone cells

1 The image of objects that we are looking directly at (i.e. which are in the centre of our field of vision) falls on an especially sensitive part of the retina — the **fovea** (or yellow spot). This region has far more cones than rods. There are **three** types of **cone** cell and each of the types is sensitive to one type of light: red, green or blue. Colour-blind people are deficient usually in the red and green cones. As well as being responsible for **colour vision**, cone cells also give a sharper image than rod cells, while **rod** cells are more **light-sensitive** than cone cells, and are thus more efficient than cones in **dim** light.

TIP

Cones and Colour vision both begin with a C.

2   There are no rods or cones at the point where the retina is joined to the optic nerve. Images formed on this part of the retina are not converted into impulses and relayed to the brain. This region is therefore called the **blind spot**. We are not (usually) aware of the blind spot since the blind spots in our two eyes do not record the same part of our field of view and one eye thus covers for the other.

Progress check 13.3

1   It is easier to see the outline of an object in dim light by looking to the side of it. Can you explain why this is?

2   Why do you think that a 'round-arm', punch aimed at the side of a boxer's face is often more successful than one aimed straight at him?

Accommodation

Accommodation is the ability to change the focal length of the lens so that we can see both near and distant objects. Accommodation depends on:

- the elasticity of the lens

- the existence of ciliary muscles, which are used to alter the shape of the lens

- the suspensory ligaments, which transfer the effect of the ciliary muscles to the lens.

Near and distant objects

When a person is viewing a near object:

1   The (circular) ciliary muscles contract reducing their circumference.

2   They reduce pull on the (elastic) suspensory ligaments.

3   With less force on the lens, its elasticity allows it to become wider (bulge) – decreasing its focal length and refracting light rays to a greater degree.

4   Rays from the near object produce a focused image on the retina.

When a person is viewing a distant object:

1   The (circular) ciliary muscles relax, increasing their circumference.

2   The suspensory ligaments are pulled tight.

3   The lens is stretched to become longer and thinner – increasing its focal length and decreasing the amount that the light is refracted. This is helped by muscles at the back of the eyeball that contract to increase the liquid pressure in the eye.

4   Rays from the distant object are brought to focus on the retina.

The pupil (or iris) reflex

This **reflex** is usually described as the pupil reflex but, since the pupil is a space, it is the muscles of the iris that respond. Therefore, it is more accurate to call it the iris reflex.

Bright light could seriously damage the delicate light sensitive cells of the retina. The intensity of light falling on the retina is therefore controlled by the iris. In general terms, the size of the pupil automatically adjusts to the intensity of light reaching the eye. In bright light, the iris increases in size to make the pupil small, reducing the light that reaches the retina. In dim light, the iris opens up the pupil to allow as much light into the eye as possible.

How the iris muscles work to control pupil size

The iris has an antagonistic arrangement of circular and radial muscles.

In bright light

1   Light sensitive cells in the retina detect the light intensity.

2   Impulses are sent along the optic (a sensory) nerve to the brain.

3   The brain returns impulses along a motor nerve to the circular muscles of the iris.

4   The circular iris muscles contract while the radial iris muscles relax.

5   The diameter of the pupil, the hole in the centre, decreases, allowing less light to enter, decreasing the risk of damage to the retina.

In dim light

The reverse of the bright light process occurs:

1 The radial iris muscles contract.

2 The circular iris muscles relax.

3 Increase of the diameter of the pupil allowing more light to enter.

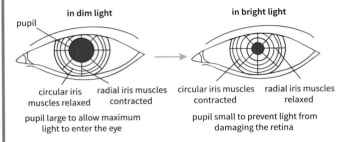

Figure 13.7 Antagonistic muscles in the iris

> **TIP**
>
> It is essential that you know the difference between the iris muscles and the ciliary muscles and how they achieve their particular functions.

Progress check 13.4

1 Where does the process of accommodation occur in the eye (Figure 13.8)?

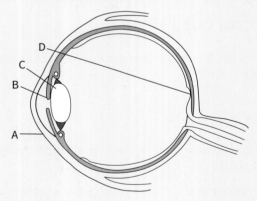

Figure 13.8

2 Which is the effector in the iris (pupil) reflex?

A   a motor neurone

B   a sensory neurone

C   the iris muscles

D   the yellow spot (fovea)

Worked example

With reference only to labelled regions on Figure 13.8, explain how, in bright light, a person is able to see an object directly in front of them.

Answer

First, note that the answer should be restricted only to the regions A, B, C and D. Also note that the words 'bright light' in the question indicates that there needs to be a description of the effect of bright light on the eye and thus your answer will have two parts: (i) the effect of bright light on the eye and (ii) how the eye enables you to see the object. Deal first with the effect of light: since bright light might damage cells in the retina, the amount of light has to be restricted. This is achieved by the circular muscles of the iris (B) contracting while its radial muscles relax. This narrows the pupil, reducing the amount of light passing into the rest of the eye.

In order to see the object, light rays from the object are first refracted (converged) as they pass through the transparent cornea (A). To ensure that a sharp image is formed on the retina, the focal length of the elastic lens (C) is adjusted (fatter with a shorter focal length for objects that are nearer to the person). Since the object is directly in front of the person, its image will be formed on the fovea (D) of the retina. This will ensure an image that is in the sharpest focus and in the best colours, since the sensitive cells responsible for this (the cones) are found in the greatest concentration in the fovea. The cones send impulses to the optic nerve and the brain so that the image is seen.

Note that as much information as possible is packed into the answer without straying into description of other structures in the eye. For example, the iris muscles are identified (circular and radial), the cornea is described as transparent and the lens as elastic.

13.03 Hormones in humans

A **hormone** is a **chemical substance, produced by a gland and carried by the blood, which alters the activity of one or more specific target organs**.

Hormones must be relatively small, soluble, easily diffusible molecules, so that they can pass quickly from the cells that make them into (and later out of) blood capillaries.

Since the **endocrine** glands that produce hormones pass them directly into the blood, they have no ducts and are called **ductless glands**.

Endocrine (or ductless) glands of the human body include:

- The adrenal glands that secrete the hormone adrenaline.

- The **pancreas** that secretes the hormone **insulin** (which changes excess glucose into glycogen, stores it in the liver and makes cell membranes more permeable to glucose) and also to the hormone glucagon.

- The **testes** that secrete the hormone testosterone (responsible for the male secondary sexual characteristics – see section on sex hormones in humans in Chapter 17).

- The **ovaries** that secrete, amongst others, the hormone oestrogen (responsible for the female secondary sexual characteristics – see section on sex hormones in humans in Chapter 17).

The adrenal glands

The main features of the adrenal glands are given in Table 13.1.

Adrenaline is produced in situations such as the moments before the start of a competitive event (including an examination!), when being chased by an angry dog or during a heated argument. Some other effects of adrenaline are increasing blood pressure, diverting blood away from the intestines and towards the muscles about to be used and increasing air flow to the lungs. All these effects help to make the body work more efficiently to meet the emergency.

 TIP It is common for students to confuse hormones with enzymes. Make sure that you know the difference.

Adrenaline and metabolic activity

An increase in adrenaline in the blood increases the metabolic rate of the body. As the metabolic reactions within the cells are occurring at a faster rate, they must be supplied with more oxygen (deeper breathing) and more glucose (adrenaline converts glycogen stored in the liver into glucose, which is released into the blood and travels to the cells). Carbon dioxide also has to be removed more quickly – helped by the deeper breathing. In order that all these processes can occur, the heart beats more quickly and, under the effect of adrenaline, sends out more blood with each contraction.

Comparison of nervous and hormonal control

Our nervous system and our hormones work together to coordinate our body responses and activities. They have some similarities:

- They both require a stimulus.

- They both produce a response.

- The intensity of the stimulus determines the scale of response.

However, they work in rather different ways as shown in Table 13.2.

Gland	Where situated	Hormone produced	Target organs	Effect of hormone
adrenals	above the kidneys	adrenaline (the 'fight, fright and flight' hormone)	muscles that control breathing	deeper breaths – more oxygen to brain and muscles
			heart	heart beats faster (faster pulse rate) – more oxygen to brain and muscles
			eyes (iris muscle)	widens pupils

Table 13.1 Main feature of the adrenal glands

Nervous control	Hormonal control
impulses travel along neurones	chemicals travel in the blood
transmission very rapid	transmission much slower – at the speed of blood flow
very rapid response	delayed response
a specific effector responds	responses often widespread
response quick and short-lived	response often prolonged

Table 13.2 Differences between nervous and hormonal control

Chapter summary

- You have learnt the structure of neurones and how neurones are involved in reflex actions.

- You now know what hormones are and know, in detail, the effects of adrenaline.

- You have seen how the nervous and endocrine systems differ in the way they work.

- Students following the supplementary course have learnt the details of accommodation and the iris reflex.

Exam-style questions

1   Figure 13.9 shows the structures involved in a reflex arc.

   a   On Figure 13.9, label i) the receptor; ii) the effector; iii) the relay neurone. [3]

   b   Name two types of tissue that can act as effectors. [2]

   c   Describe how the structures shown in Figure 13.8 are involved in a reflex action. [6]

[Total 11]

2   Describe the events that occur from the time an impulse arrives at the end of one neurone to the time an impulse passes along the next neurone. [7]

3   Explain why feeling nervous before a race can help to improve an athlete's performance. [10]

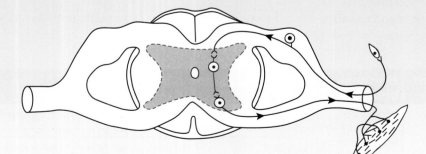

Figure 13.9

Homeostasis

14.01 Homeostasis

The body will operate at its most efficient only if conditions within it remain (more or less) stable. As has been shown, the kidneys have the important function of keeping the blood plasma at a constant concentration. Thus the tissue fluid bathing the cells is always at a constant concentration and, since cell membranes are partially permeable, osmosis will ensure that the concentration within all cells is also always constant. Many organs within the body play a part in maintaining constant conditions and are described as organs of homeostasis.

Homeostasis is defined as **the maintenance of a constant internal environment**.

The concept of control by negative feedback

In order for a constant internal environment to be maintained, it is necessary to have receptors (sensors) in the body capable of detecting when the environment fluctuates too far either side of the required state. There must then be some form of mechanism that comes into operation in order to return the conditions to normal. All homeostatic systems have a mechanism

that automatically brings about a correction, no matter which side of the optimum the change has occurred. This is called a **negative feedback** system.

The control of blood glucose concentration

This is an example of negative feedback.

After a meal containing carbohydrate, the blood glucose concentration rises but immediately, special cells in the **pancreas** called the **islets of Langerhans** begin to produce the hormone insulin (Table 14.1). Under the influence of insulin:

- cells more readily absorb glucose
- soluble glucose is converted in the **liver** and **muscle cells** to insoluble **glycogen** and **stored**.

This reduces the blood glucose concentration.

If the glucose concentration in the blood falls below the level required for the body to function properly, then the **islets of Langerhans** in the pancreas secrete the hormone glucagon, which converts the stored **glycogen** in the liver back to glucose and releases it into the blood (Table 14.1).

Gland	Hormone produced	Target organs	Effect of hormone
pancreas	insulin	liver and muscles	promotes the uptake of glucose by cells and the conversion of glucose to glycogen for storage
pancreas	glucagon	liver and muscles	promotes the conversion of glycogen to glucose

Table 14.1 Control of blood glucose

Diabetes

If a person's islet cells produce insufficient insulin, then the level of glucose in their blood will rise. In children and young adults, this may happen suddenly, and is a symptom of **type 1** diabetes.

Other symptoms include:

- Glucose is present in the urine.
- The patient urinates more frequently.
- The patient becomes thirsty and tired.
- Weight loss and generally feeling unwell

Treatment will include regular injections of insulin. The carbohydrate intake must then be regulated to match the amount of insulin injected.

Progress check 14.1

1   Name a hormone that is responsible for raising blood glucose concentration.

2   Name the carbohydrate that is stored in the liver.

The skin

The skin is the largest organ of the human body (Figure 14.1). It is an important sense organ but also forms the barrier between the body and the external environment. It is the organ through which we may both gain and lose heat. This is particularly important in the homeostatic process of **temperature regulation**.

The part played by the skin in temperature regulation

When the temperature of the blood rises above 37 °C :

1   The sweat glands release more sweat.

2   The sweat evaporates removes heat from the skin as it does so.

Thus more heat is lost to the environment, cooling the blood and allowing the body temperature to return to normal.

NB The person may also take cold drinks and seek shade.

When the temperature of the blood falls below 37 °C:

1   Sweating is greatly reduced.

2   The fatty tissue acts as an insulator.

Thus less heat is lost to the environment, allowing metabolic processes to release energy in the form of heat to help bring the body temperature back to normal.

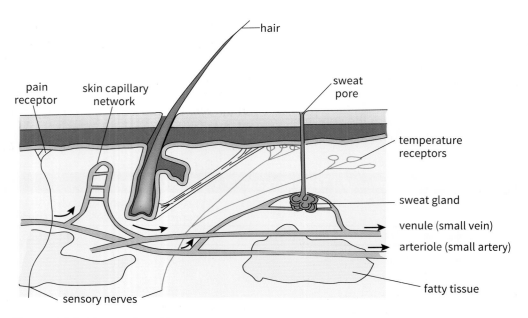

Figure 14.1 Mammalian skin

Shivering

If these measures prove inadequate, muscles in the body will start to contract and relax rhythmically, it is called shivering. It releases heat energy which increases the body temperature.

NB A person may take warm drinks, increase the insulating layer of air around the body by wearing more clothes and do some exercise, all of which will help to raise body temperature.

Temperature receptors in the skin and brain

As a sense organ, one of the properties of the skin is to be able to detect temperature change in the external environment. A change in atmospheric temperature will thus start some of the temperature control mechanisms, even before there is any appreciable change in the temperature of the blood. The temperature change is detected by temperature receptors in the skin and impulses are sent to the central nervous system by sensory neurones. A reflex action sends impulses through motor neurones to effectors, including sweat glands, in the skin. Changes in the blood temperature are detected by the hypothalamus situated on the under-side of the brain. This is the structure responsible for the coordination of all the temperature control mechanisms.

Progress check 14.2

1 Name the term used for the maintenance of a constant internal environment.

2 State three ways the body counteracts heat loss.

The importance of the blood vessels in the skin during temperature regulation

It is essential that body cells are kept at a temperature within narrow, set limits that allow enzymes controlling metabolism to work at their optimum rate. Thus the ability to use the skin to lose or conserve heat is of great importance. Central to this function is the capillary network present in the skin. The more blood present in the capillaries, the more heat is being carried to the skin surface, as blood is the carrier of heat round the body.

Capillary walls are only one cell thick. They have no muscles and thus cannot actively increase or decrease their diameter, or, therefore the amount of blood in them. This is done by the blood vessels that supply them with blood – the arterioles (small arteries).

 TIP Capillaries never rise up nearer the skin surface when the person is hot, or sink further from the skin surface when they are cold.

Vasodilation

When heat needs to be lost, the arterioles in the skin dilate – that is, their muscular walls relax and the arterioles widen, increasing the diameter of their lumens allowing more blood to flow. More blood enters the capillaries and they widen accordingly. More blood flows to the sweat glands and more sweat is extracted from the blood and passed up to the skin surface to increase evaporation.

Vasoconstriction

When the body temperature falls and heat needs to be retained, the muscles in the arteriole walls contract and thus, the arterioles constrict (become narrower). Less blood is therefore sent to the capillaries. A capillary network often has a by-pass or shunt vessel. The blood vessel that supplies the capillaries divides into two, and the branch leading to the capillaries has a ring of muscle at its base. In order to retain heat, this muscle contracts and closes off the loop to the capillaries. Instead of passing to the capillary network, the blood is sent through the shunt vessel, which later re-joins the vessel from the capillaries. Vasoconstriction also results in less blood passing to the sweat glands. Less sweat is poured onto the skin surface to evaporate and further cool the skin.

How the regulation of body temperature illustrates negative feedback

The way body temperature is regulated by negative feedback is shown in Figure 14.2, detected by the hypothalamus in the brain

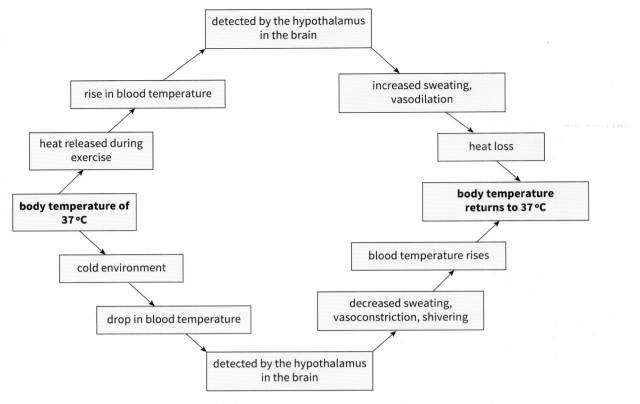

Figure 14.2 Body temperature regulation

Worked example

Can you suggest why an oven set at 170 °C is an example of a negative feedback system?

Answer

To answer this question, you need to have the basic knowledge that an oven is heated by an element and that the thermostat is connected to a power switch. If you were not aware of this, then the preceding sentence would provide you with the necessary information. Setting the thermostat to 170 °C is equivalent to fixing the norm or set point. The heating element is the input into the system. The thermostat is the sensor. When the temperature reaches just above the set point, the thermostat switches off the power to the element. Heat gradually escapes from the oven until the thermostat (sensor) detects that the temperature has fallen below the set point when it switches the element back on again. Thus there is a system that automatically brings about a correction whenever there is a fluctuation either side of the set point.

14.02 Tropic responses

Plants show **sensitivity** (one of the characteristic of living organs) when their organs **grow** in a particular direction in response to stimuli. Two of those stimuli are **gravity** and **light**.

A response in which a plant grows towards or away from gravity is known as gravitropism (or **geotropism**).

A response in which a plant grows towards or away from the direction from which light is coming is known as phototropism.

TIP

Make sure that you spell 'tropic' correctly, since 'trophic' is to do with feeding, not with response to stimuli.

PRACTICAL

Aim: **To show how plants shoots and roots respond to gravity**.

Apparatus: A wide-mouthed jar

A piece of thick cardboard, just wider than the mouth of the jar

A polythene sheet the same size as the piece of cardboard

Cotton wool

A petri dish

A pin

3 suitable large seeds (e.g. broad bean)

Method:

Spread a thin layer of cotton wool in the bottom of the petri dish. Place the seeds on the cotton wool (evenly spaced). Pour water onto the cotton wool (but do not cover the seeds). In order to reduce evaporation, place the lid on the petri dish (do not apply force) and leave in a warm room until the seeds have germinated (24–48 hours, depending on temperature). The cotton wool should be checked regularly to ensure it has not dried out. Select the seedling with the straightest young root (radicle) and young shoot (plumule).

Pour some water into the jar – to a depth of about 1 cm – and using a little of the wet cotton wool, use the pin to fix the selected seed to the cardboard as shown in Figure 14.3. The polythene will keep the cardboard dry, the water will keep the air in the jar humid – reducing evaporation from the seed and cotton wool. Place the set-up in a dark cupboard or cover with black polythene. Examine the results every 24 hours.

Results:

The radicle (young root) will be seen to have grown downwards towards gravity and the plumule (young shoot) will be growing upwards, away from gravity.

The radicle is said to be showing positive (+) gravitropism. The plumule is showing negative (−) gravitropism.

NB A piece of apparatus called a clinostat has a platform that revolves four times per hour. If the platform is arranged horizontally, and the seedling is pinned to the platform with its radicle or plumule pointing horizontally from the platform, it will continue to grow horizontally whilst the platform rotates. The rotation ensures that gravity acts equally on all sides of the plant organs, eliminating its effect on their growth. The clinostat thus provides a control for the experiment.

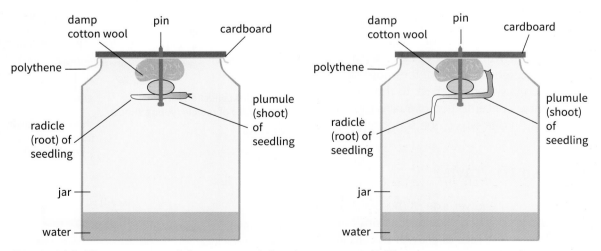

Figure 14.3 The response of the roots and the shoots to gravity. The roots grow towards gravity (+ geo- or gravitropism). The shoots grow away from gravity (− geo- or gravitropism)

Aim: **To show how plant shoots respond to light arriving from one side. (This response is called phototropism.)**

Apparatus: 2 small plant pots (e.g. small yoghurt pots cut to about 5 cm high) containing damp soil

2 cardboard boxes (with lids) about 15 cm wide

A bell jar

A large beaker or transparent cover of approximately the same size as the boxes

A sharp knife

Razor blade or scalpel

Several seeds – all of the same species (e.g. cress, mustard)

Method:

Place three or four of the seeds in the damp soil and leave to germinate in a dark cupboard. In the lid of one of the boxes, cut a narrow slit about 5 cm long and 5 mm wide. The slit should be in the middle of the lid and parallel to its long axis. Stand the two boxes on end, and when the seeds have germinated, and

their plumules are visible above the soil, place one of the pots in each of the two boxes and replace their lids. Place the third pot under the transparent cover and place in a well-lit location. Leave the experiment for 48 hours.

Results:

1 Seedlings under the transparent cover: light being received all round: green, healthy stems growing straight upwards.

2 Seedlings grown in the box with no light entering: tall, thin straggly but generally straight yellow stems.

3 Seedlings grown in box with light coming from one side only through the slit: shoots have grown towards the slit. They have shown positive (+) phototropism.

NB The seedlings in the box with no light and those under the transparent cover provide controls for the experiment.

TIP

Never say that a root or shoot 'bends' towards or away from a stimulus. The response is a growth movement – they grow towards or away from stimuli.

An explanation of tropisms

In plants, the very tip (apex) of shoots and roots have cells (receptor cells) specialised to receive stimuli. The elongating cells that occur just beyond (or behind) the region of cell division are the **effector** cells and the link between the two is growth hormones called **auxins**. Auxins are manufactured at the tip of the shoot or root and, in solution in water, pass back to the growth regions. A **differential** concentration of auxins causes a **differential growth** on one side of the shoot or root.

The response to gravity

Auxins accumulate on the lower side of (horizontal) shoots and roots due to gravity. However, **auxins** have the **opposite effect** on the **rate of growth of cells** in the root to that on cells of the shoot. In **roots**, they **slow down** the rate of growth. In **shoots**, they **speed up** the rate of growth.

Thus, a horizontal root grows downwards (positive [+] geotropism) and a horizontal shoot grows upwards (negative [–] geotropism) (Figure 14.4).

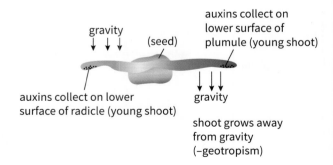

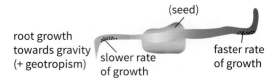

Figure 14.4 Geotropism in a seedling

The response to one-sided light

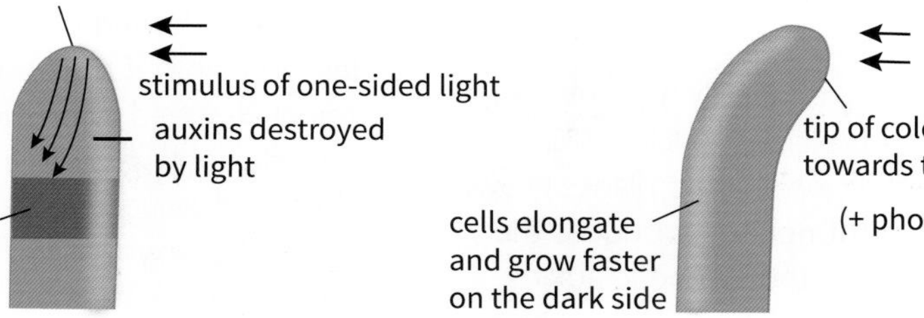

tip of coleoptile receives the stimulus and makes auxins

auxins pass back to the region of cell growth

stimulus of one-sided light

auxins destroyed by light

greater concentration of auxins on the dark side

tip of coleoptile grows towards the stimulus

(+ phototropism)

cells elongate and grow faster on the dark side

Figure 14.5 Positive phototropism in a maize coleoptile

Monocotyledonous plants (e.g. maize) are often used to demonstrate this effect, since the shoots are covered by a sheath (the coleoptile) for the first few days of growth. The coleoptile responds to stimuli.

As shown in the figure, the coleoptile responds to one-sided light by **growing** towards it. This is termed **positive [+] phototropism**. It is caused by the accumulation of auxins on the dark side of the coleoptile. In shoots, auxins stimulate the growth of cells, thus those on the darker side grow (elongate) faster than those on the illuminated side.

NB All the cells in the growth regions of the shoots in the dark and those in all-round light receive equal quantities of auxins. Thus all cells grow at the same rate, and the stems grow straight. Light is thought to have a destructive effect on auxins, explaining why those in the dark grow longer than those in the light. The yellow leaves in the dark are due to the fact that green chlorophyll cannot be manufactured in the absence of light. Long straggly, yellow stems bearing yellow leaves are said to show a condition called **etiolation**.

Progress check 14.3

I Figure 14.6 shows a growing plumule. From point P in its growth, from which directions did it receive light?

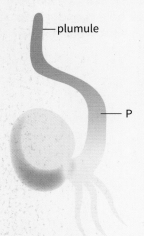

— plumule

— P

Figure 14.6

A first from above and then from all round

B first from the left and then from all round

C first from the left and then from the right

D first from the right and then from above

2 What is the effect of auxins on the rate of growth of:

i) cells in a seedling's radicle and

ii) cells in a seedling's plumule?

The use of synthetic plant hormones as weed killers (herbicides)

Plant hormones are often called 'plant growth substances'.

Auxins can be manufactured synthetically and relatively cheaply and can be used as herbicides to kill weeds. When a herbicide is applied to a plant, the plant's growth is stimulated. However, the plant neither produces the enzymes needed to break down the synthetic auxins,

nor is it able to produce food fast enough through photosynthesis to sustain its growth rate.

Selective hormone weed killers are designed to kill only certain types of plants. For example, many weeds are **broad-leaved** plants, while cereal crops are narrow-leaved plants. If a crop in a field is sprayed with a selective weed killer such as **2,4-D** that causes rapid and unsustainable growth in the weeds and is effective only on broad-leaved plants (**dicotyledonous plants**), then most of the weeds will die, leaving the narrow leaved crop plants unharmed.

Chapter summary

- [ ] You have learnt what homeostasis is and some examples of it in the human body.

- [ ] You have learnt how plants respond to external stimuli.

- [ ] You have also conducted experiments to demonstrate the responses of roots and shoots to external stimuli.

- [ ] Students following the supplementary course have learnt the mechanism within roots and shoots that bring about these responses.

Exam-style questions

1 Figure 14.7 shows a diagram of human skin with structures A and B labelled.

   Describe how a constant body temperature is maintained in humans. Your answer should include the identification of structures A and B and an account of the part they play in the process. [7]

2 Explain why the process of maintaining a constant body temperature illustrates control by negative feedback. [8]

3 Two samples of watered, germinating seedlings, A and B, are placed in a black box. The shoots of seedlings A have had the tips of their plumules cut off. The box has a hole in one side to allow light to enter. Explain what would happen after 2 days in the box. [10]

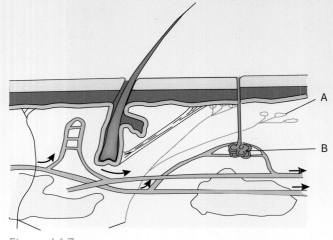

Figure 14.7

Drugs

Learning outcomes

By the end of this chapter, you should understand:

☐ How drugs are used as useful medicines

☐ How drugs can be misused and the harm that may cause

15.01 Drugs

Hormones are chemicals manufactured by and within a living organism, which have a specific effect on that organism's metabolism. Chemicals, not made by the organism, may be introduced into the body of the organism for a specific effect. Such chemicals are called drugs.

A **drug** is defined as **any substance taken into the body that modifies or affects chemical reactions in the body**.

Medicinal drugs

Drugs may be used for beneficial effects:

1   Pain relief: for example, aspirin, paracetamol and morphine.

2   Treatment of bacterial infections: diseases caused by bacteria, such as syphilis are treated with drugs called antibiotics, of which a well-known example is penicillin. **Viruses** are **unaffected** by **antibiotics**. There are many different types of antibiotics and the antibiotic is usually accurately matched to the type of bacterium causing the disease. 'Broad spectrum' antibiotics are used to kill a wide range of bacteria and may be given when the exact type of bacteria has not been identified. 'Narrow spectrum' antibiotics are effective against only a very few types of bacteria. Bacteria are generally becoming more resistant to antibiotics. Effective antibiotics are becoming difficult to find even for some bacterial infections that used to be successfully controlled by antibiotics.

15.02 The dangers of misuse and overuse of antibiotics

The complete prescribed course of antibiotics must be taken – even if the patient is feeling better. In this way all the pathogenic bacteria are killed. Ending the treatment before the end of the course allows the most resistant bacteria to remain alive and perhaps be passed on to other people who then find that the antibiotic is far less effective.

The over-prescription of antibiotics has a similar effect, but it may take longer for the problem to appear. A few resistant bacteria may remain alive even when the full course is taken. These resistant bacteria then breed further resistant forms that are passed on to other people. There is now a trend not to use antibiotics if the infection is not a serious one, leaving the body to develop its own resistance and immunity to the bacterium.

A type of bacterium that infects open wounds has become particularly troublesome in hospitals as it has become resistant to a wide variety of antibiotics and is therefore difficult to treat. It was named from the fact that the bacterium responsible for the condition is called *Staphylococcus aureus* and, because it was first found to be resistant to the antibiotic Methicillin and later found to be resistant to many drugs, it has been given the name MRSA (Multidrug- or Methicillin-Resistant *Staphylococcus aureus*). Tuberculosis is a disease that is treated with antibiotics, but the causal bacterium has developed resistance to the antibiotics that were once used very successfully to treat the disease.

Antibiotics are effective only against bacteria – not against viruses

Antibiotics are effective against bacteria and not against viruses since they interfere with various metabolic processes of the bacteria.

- They may **stop** bacteria from making new **cell walls** (e.g. penicillin). This is the process employed by the fungus *Penicillium* that releases the chemical to prevent the growth of bacteria that would otherwise represent competition for the substrate both organisms might be attempting to feed on.

- Some (e.g. tetracycline) inhibit the production of protein in the bacterial cell – therefore it cannot grow.

- Some prevent bacteria from undergoing cell division.

- Others block the chemical reactions that are part of the process of respiration thus inhibiting energy release.

Because none of the processes mentioned here occur in viruses, antibiotics are ineffective against them.

Progress check 15.1

1   A bacterial cell wall is a non-living structure enclosing a cell. Why should an antibiotic that prevents bacteria from making new cell walls kill the bacteria?

2   Why has MRSA become a problem in hospitals?

15.03 Misused drugs

If a drug is taken simply for enjoyment (sometimes called a 'recreational' drug), and often therefore in high dosage, it can lead to **addiction**.

Heroin

Heroin is a drug that is abused in this way. It is a powerful depressant (i.e. it depresses the function or activity of nervous system – particularly the brain).

Heroin affects the parts of the brain that control pain, breathing and blood pressure. It produces a feeling of extreme well-being and relaxation, which removes all feelings of anxiety. (See also how it affects transmission at synapses – Chapter 13.)

It is a drug to which the body shows tolerance. As a result progressively increased dosages are required in order to give the same effect.

This leads to a state of dependence in which the user cannot face life without the drug. The user begins to crave the drug, and if unable to obtain further supplies, suffers severe withdrawal symptoms that include diarrhoea, vomiting, muscular pain, shaking and hallucination. Abuse of the drug has led to addiction.

Addiction can lead the user into a life of crime in order to obtain money and/or regular supplies of the drug and, since heroin is a drug normally taken by injection into a vein, the use by several addicts of the same unsterilised needles may result in the transmission of blood-borne diseases such as hepatitis and HIV/AIDS.

Alcohol

Alcohol is a more widely available and more widely used drug. It is generally regarded as a more 'socially acceptable' drug by many societies, but its effects are as follows:

1   It is a depressant, creating a feeling of well-being.

2   It slows reaction times (increasing the risk of accident).

3   When consumed in excessive quantities it leads to loss of self-control. Many people under the influence of alcohol behave in a way in which they would be ashamed of when sober.

4   Alcohol and other toxins are broken down in the liver (which detoxifies them) but continuous and excessive detoxification leads to liver damage (cirrhosis), which can eventually prove fatal.

Like heroin, alcohol is a drug of addiction and in order to prevent **withdrawal symptoms** (which are not as severe as those for heroin, but include mental confusion, shaking and unconsciousness) considerable amounts of money are spent to satisfy the cravings. Often a person's family suffers, not only from a lack of financial support, but also from physical violence that often accompanies alcoholism.

Nicotine

Nicotine is the drug of addiction present in **cigarette smoke**. A person suffers (relatively) mild withdrawal symptoms if the craving is not satisfied. Nicotine has the following effects:

1   It is a poison that increases heart rate and blood pressure.

2 It may cause blood clotting increasing the risk of thrombosis (which, if in the coronary artery of the heart will result in a **heart attack**).

3 Its withdrawal symptoms encourage the person to continue smoking.

Other harmful components of cigarette smoke

Tar forms a layer over the walls of the alveoli. As a result, it restricts gaseous exchange. It is also a carcinogen and prolonged exposure to it may lead to **lung cancer**.

Carbon monoxide is taken up permanently by hemoglobin in preference to oxygen, forming carboxyhemoglobin. It greatly reduces the ability of the blood to carry oxygen.

Irritant chemicals and particles cause cells lining the bronchi and bronchioles to increase their production of mucus. These chemicals also destroy the cilia lining the trachea, which sweep away the dirt in a 'moving carpet' of mucus and carry it to the throat for swallowing. The build-up of mucus is relieved only by continual coughing (smoker's cough). Persistent coughing eventually damages the walls of the alveoli (emphysema). The airways may become narrower with time restricting the airflow into and out of the lungs. This condition is known as **chronic obstructive pulmonary disease (COPD)**.

A pregnant woman who smokes also risks the heath of her baby. Less oxygen reaches the baby as a result of the effects of carbon monoxide and nicotine can pass from the mother's to the baby's blood. Babies born to mothers who smoke during pregnancy have been shown to be underweight and, perhaps less intelligent, and there is a greater risk of miscarriage.

Passive smoking: Evidence now exists that breathing the smoke from other people can be harmful and cigarette smoke is certainly an irritant to the eyes and leaves a lingering smell in clothes. Smoking is therefore increasingly becoming a socially unacceptable habit.

Progress check 15.2

1 What is meant by 'tolerance' with used in connection to a drug?

2 Which component of cigarette smoke is most likely to cause blood to clot?

A carbon monoxide

B irritant chemicals

C nicotine

D tar

The link between smoking and lung cancer

There have been many research projects on the link between smoking and lung cancer.

Studies have shown that a person who smokes for 45 years has a 100 times greater risk of lung cancer than a person who smokes for 15 years.

Smoking one packet of cigarettes a day for 40 years is eight times more dangerous than smoking two packets a day for 20 years.

Even light or irregular smoking can increase the risk of cancer. People who smoke one to four cigarettes a day have a much greater risk of dying from lung cancer or heart disease. People who smoke just two cigarettes a day are more likely to develop cancers of the mouth and oesophagus.

And smoking is linked to other cancers too – occasional smokers who have never smoked daily still have higher risks of most cancers, including double the risk of bladder cancer.

Worked example

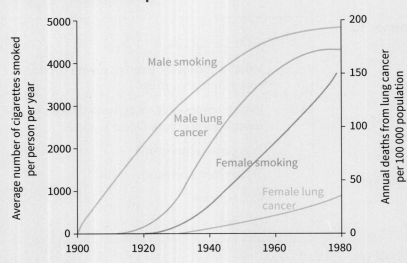

Figure 15.1

a Explain why nicotine in cigarette smoke is dangerous to human health.

b The graph show the level of smoking and the incidence of lung cancer in one country from 1900 to 1980.

   i) Name the substance in cigarette smoke that is the major cause of lung cancer.

   ii) Suggest reasons for deaths from lung cancer in females differing from those for males in the year 1960.

Answer

a There is often the mistaken belief that it is nicotine that causes lung cancer. Its main harmful effects are to increase the tendency for blood to clot in the blood vessels. If this happens in the coronary artery it may lead to a heart attack, if it happens in blood capillaries in the brain, it can cause a stroke. Nicotine also tends to increase blood pressure and make the heart beat faster, which, together, can cause circulatory disorders and damage to other organs (e.g. the kidneys). The 'hidden' danger, however, is that it is a drug of addiction. Its absence leads to withdrawal symptoms that encourage the smoker to light up another cigarette thus continuing to increase the chances of harmful effects.

b i) Tar (which coats the surface of the respiratory organs – particularly the alveoli).

   ii) The question says only that they 'differ'. You need to state how – and you should also include readings from the graph. In women, in 1960, there were about 25 deaths per 100 000 (*always* remember to include units), whereas in men, there were 150 deaths per 100 000 (i.e. 6 times more than in women). The graph indicates that a reason for this is that women were smoking approximately 2000 cigarettes per year compared with men who were smoking approximately 4500 per year. Women were thus not being exposed to such high levels of tar, but also the graphs show that the number of related deaths lag by just under 20 years behind the onset of smoking. It would also appear that deaths from lung cancer in women is not so clearly associated with smoking as it is in men. Women started smoking about 20 years later than men and the numbers smoking were rising at the same rate as they did for men, but the number of lung cancer deaths was rising much more slowly.

Hormones used to improve athletic performance

Other abuse of drugs includes their use to enhance sporting performance. Athletes discovered doing so can expect to receive a life-long ban from sport. The drugs commonly involved are:

- **Testosterone:** which, as a hormone that naturally builds muscle in a male at puberty and maintains it in adulthood, is used to increase muscle development.

- **Other anabolic steroids:** These have effects similar to testosterone. They are commercially made though they do occur naturally in cells. Anabolism is that part of metabolism the deals with the building up of chemical molecules and thus tissues. When those tissues are muscles, then there will often be an improvement in muscular performance.

However, there are long-term disadvantages to the use of testosterone and steroids. Steroids, in particular, can cause problems such as high blood pressure, dangerous levels of cholesterol in the blood and liver damage.

Chapter summary

- ■ You have learnt about the use of medicinal drugs including antibiotics.

- ■ You have also learnt about drug misuse.

- ■ You now know what is meant by addiction, tolerance and withdrawal symptoms.

- ■ You also know the harmful effects of each of the main components in cigarette smoke.

Exam-style questions

1 Figure 15.2 shows the antibiotic resistance rates of pathogens for three antibiotics.

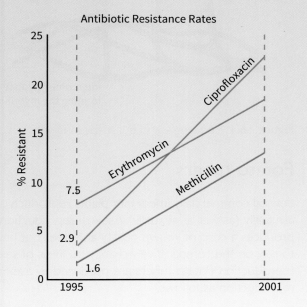

Antibiotic Resistance Rates

Figure 15.2

a   i)   State the type of pathogen against which antibiotics are used. [1]

ii)   State the conclusions that can be drawn from the graph in Figure 15.2 about the effectiveness of the three antibiotics between 1995 and 2001. [4]

iii)   State which of these antibiotics would be most likely to be termed a 'broad spectrum' antibiotic. [1]

[Total 6]

2 Explain the possible causes of the results shown in Figure 15.2. [6]

3 a   Explain the short-term harmful effects of alcohol. [5]

   b   Explain how a person, and the society in which they live, can be adversely affected by their long-term use of alcohol. [5]

[Total 10]

Reproduction in plants

16.01 Reproduction

There are two methods by which organisms can increase (or maintain) their numbers. They are **asexual** and **sexual reproduction**.

Asexual reproduction

Asexual reproduction is defined as **the process resulting in the production of genetically identical offspring from one parent**.

Asexual reproduction is common in simple organisms such as bacteria and fungi, in many plants and in several animals (but not mammals). In all cases, it involves **cell division** almost always by a process called **mitosis**. During cell division the cell's DNA produces an identical copy of itself, with each of the new cells formed receiving one of the copies.

Bacteria

Asexual cell reproduction begins as soon as a bacterial cell lands on a suitable substrate. It will continue so long as there is a supply of food, the temperature is suitable, oxygen is present (if the bacterium in question respires aerobically) and there are no poisons to inhibit the cell's metabolism. In suitable conditions, a bacterial cell may divide every half an hour.

Fungi

After feeding and growing on a suitable substrate, fungi produce 'fruiting bodies' that project from the substrate. These bodies are called sporangia, inside which the cells divide to form many identical **spores** (Figure 16.1). Eventually, the sporangium bursts and the small, light spores are carried away by air currents. When they land on a suitable substrate, they germinate to produce a new mycelium.

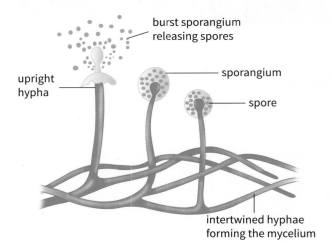

Figure 16.1 Sporangia and burst sporangium

Potato tubers

Several commercially important plants reproduce asexually (as well as sexually). Asexual reproduction produces a crop of known quality and does not have to rely on the comparatively chancy business of sexual reproduction. Potato tubers are produced by asexual reproduction as follows.

The parent potato plant grows branch stems at, or just below, ground level (Figure 16.2). The ends of these branch stems swell up to form bulbous structures called (stem) tubers. These are used to **store starch** – in order to keep the plants alive until the next growing season. The parent plant then dies, leaving its tubers in the soil.

Being modified stems, the tubers possess **dormant buds**. When the next growing season arrives, the dormant buds begin to grow into new plants. Since the original parent will have produced many tubers, many new plants will grow – thus reproduction has occurred and all the new plants will be genetically identical to the original parent and to each other.

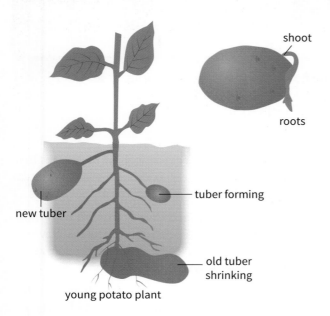

Figure 16.2 Potato plant with tubers and a sprouting tuber

A population of organisms thus produced are known as **clones**. All individuals in a clone are genetically identical.

The advantages and disadvantages to a species of asexual reproduction

The various advantages and disadvantages to plants of asexual reproduction are listed in Table 16.1.

Advantages	Disadvantages
Only one parent organism is required.	A lack of variation in offspring prevents evolution.
It is a relatively certain method of reproduction.	Adverse conditions and disease will be likely to affect all members of the population.
Since the parent can survive in that particular habitat, the genetically identical offspring are bound to be suited to the environment as well.	Overcrowding and thus competition for resources (water, nutrients and, important in plants, light).
	Except where spores are involved, distribution of the species is likely to be limited.

Table 16.1

Asexual reproduction is often used in **agriculture** to produce plants (or crops) of known quality or yield. Many cuttings from the stems or sometimes the leaves of one parent plant may be taken (**tissue culture**). The cuttings may then be dusted with rooting hormone powder to encourage root production before being planted in a medium rich in required nutrients. However, the advantages and disadvantages of asexual reproduction listed previously will also apply to commercial crops. It may prove very expensive to the farmer who grows a commercially advantageous crop that then has to be treated with insecticides and fungicides to protect that crop against disease.

16.02 Sexual reproduction

Sexual reproduction is the fusion of the nuclei of two sex cells (gametes) to form a zygote and the production of offspring that are genetically different from each other.

The **fusion of the gamete nuclei that occurs in sexual reproduction** is known as **fertilisation**.

In the body cells of most of the higher organisms, chromosomes are normally found in matching pairs – described as the **diploid** condition. Cells used in sexual reproduction (**gametes**) contain only one of each of these pairs, and are described as **haploid**. When two gametes **fuse**, the full number of chromosomes is restored to form a diploid zygote.

Gametes are produced by a special type of nuclear division called **meiosis** (reduction division), which halves the chromosome number.

Advantages of sexual reproduction to a species

- By combining the genes of two different individuals, the offspring will always show *variety*. In this way, some will be at an advantage over their fellow offspring when surviving in their environment. Those less well adapted will die and thus not pass on their disadvantageous genes. Those that survive pass on their advantageous genes, that is, **evolution** takes place more quickly when sexual reproduction is involved.

- In animals, when courtship is involved, the stronger, healthier individuals are more likely to be chosen as potential mates.

Disadvantages of sexual reproduction

- A mate may not be found.

- Fertilisation may not be successful – and in plants, agents of pollination may not be available when the flowers are mature.

- In plants, agents of fruit or seed dispersal may not be present.

- Some individuals will inherit faulty genes from both parents and thus be at a disadvantage.

Progress check 16.1

1   What happens during sexual reproduction?

    A   two diploid gametes fuse to form a diploid zygote

    B   two diploid gametes fuse to form a haploid zygote

    C   two haploid gametes fuse to form a diploid zygote

    D   one haploid gamete and one diploid gamete fuse to form a diploid zygote

2   What are the disadvantages to a species of asexual reproduction?

Agricultural applications of sexual reproduction

Advantages

By carefully choosing the organisms from which to breed (**'artificial selection'**) and because the offspring from sexual reproduction are always different from their parents as well as from one another, some of the offspring produced are always likely to be more commercially valuable even than their parents and can be used again for breeding with further advantageous results.

Disadvantages

'Inbreeding' can result in the production of weak offspring that have, or are susceptible to, disease.

16.03 Sexual reproduction in plants

TIP

Be sure not to talk of 'flowers' when you really mean 'plants'. The flowers are the reproductive organs of the plant that, depending on the type or pollination, may, or may not be colourful.

The **flowers** of plants are the organs of **sexual** reproduction but in some monocots are colourful and act like petals to attract insects.

Flowers have the following parts:

Sepals: These are (usually) green, leaf-like structures that protect the flower when it is in bud, but in some monocots they are colourful and act like petals to attract insects.

Petals: These may be large, colourful, scented with lines on them (nectar guides) to attract insects if the flower is pollinated by insects, but small, green or even absent entirely if the flower is wind-pollinated.

Stamens: These are the 'male' parts of the flower. Each stamen has a stalk (the **filament**) at the end of which is the **anther**. Thus:

Anther + Filament = Stamen

Anthers contain **pollen sacs** that make and then release pollen grains, each of which contains the male gamete(s).

TIP

Pollen grains are **not** the male gametes. They contain the male gametes.

Carpels: These are the 'female' parts of the flower. Each carpel is made up of a (sticky) **stigma** (for receiving pollen during pollination). The **style** connects the stigma to the – **ovary**, in which lie the **ovules**. The ovules contain the **female gametes**.

Pollination

Pollination is the **transfer of pollen grains from the male part of the plant (the anther of the stamen) to the female part of the plant (stigma of the carpel).**

Self-pollination

Even though the type of reproduction is **sexual**, pollen grains from an anther of a flower may be transferred to the stigma of the **same flower** (or onto the stigma of **another flower** on the same plant). This is described as **self-pollination**.

- The possibility of this occurring in the same flower is sometimes reduced by the anthers releasing their pollen grains before the stigma(s) are ready to receive pollen.

- In the absence of a suitable agent of pollination, it ensures that the plant survives for another generation.

- There is a much greater chance of the offspring being produced with genetic defects.

Cross-pollination

This is when pollen grains are transferred from an anther of a flower to the stigma of a flower on a **different plant of the same species**.

- Offspring show natural variation that allows the plant species to adapt to changes in the environment and thus to evolve.

- If plants are rare and grow at considerable distance from one another, reproduction may not occur.

- There may be an absence of pollinating agents at the time that pollen is being released.

Pollination may be brought using different agents for carrying the pollen.

The two most common are **wind** and **insect**. Wind- and insect-pollinated flowers show structural differences that are adaptations to their method of pollination.

The differences between wind- and insect-pollinated flowers

	Wind-pollinated	**Insect-pollinated**
Petals	small (or absent)	large (to act as landing platform)
	drab	colourful
	no scent	scented
	nectar guides absent	nectar guides present
	nectar absent	nectar present to attract insects
Pollen grains	small	larger
	dry and dusty	sticky
	smooth-coated	rough-coated
	light	relatively heavy
	more easily carried by wind	more easily carried by insects
	vast quantities (greater wastage)	smaller quantities (much more certain)
Anthers	held outside the flower in the wind	protected within the flower where insects will touch them
	relatively large	relatively small
Stigmas	large surface area	smaller surface area
	outside flower	inside flower

An example of a wind-pollinated flower is a grass flower (Figure 16.3).

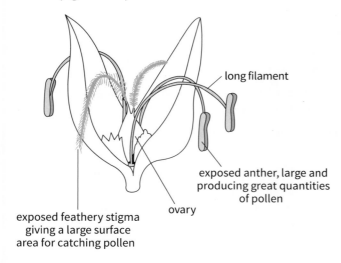

long filament

exposed anther, large and producing great quantities of pollen

ovary

exposed feathery stigma giving a large surface area for catching pollen

Figure 16.3 A wind-pollinated flower (grass flower)

The Black-eyed Susan is an example of an insect-pollinated flower (Figure 16.4).

In both wind- and insect-pollinated flowers pollen must fall on the stigma for pollination to take place (Figure 16.5).

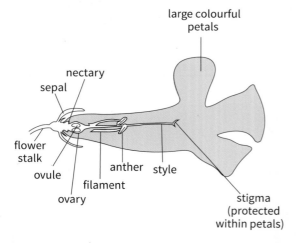

large colourful petals

nectary

sepal

flower stalk

ovule

anther   style

filament

ovary

stigma (protected within petals)

Figure 16.4 An insected-pollinated flower (Black-eyed Susan)

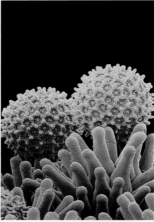

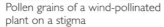

Pollen grains of a wind-pollinated plant on a stigma

Pollen grains of an insect-pollinated plant on a stigma

Figure 16.5 Pollen grains on the stigmas of a wind-pollinated plant (a) and an insect pollinated plant (b)

Progress check 16.2

1   In a developing flower, where is the pollen grain found?

　　A   anther

　　B   filament

　　C   ovary

　　D   style

2   In a developing flower, where is the ovule found?

　　A   anther

　　B   filament

　　C   ovary

　　D   style

PRACTICAL

Personal investigation of floral structure

1 A wind-pollinated flower

Use a hand lens to identify the structural features of a wind-pollinated flower (e.g. a grass). Check against the list of differences between insect- and wind-pollinated flowers previously to see how many of the differences you can observe in your two flowers.

2 An insect-pollinated flower

Using a large insect-pollinated flower, carefully remove the sepals, petals and stamens. Depending on the size of the flower, you may be able to remove the parts quite easily using your fingers, otherwise you may find a pair of forceps helpful.

Using a magnifying glass (×8 or ×10) make large drawings of one of each of the type of structure you have removed. Also, make a large drawing of the carpel still attached to the flower stalk.

Some hints on how to draw biological specimens

Drawing specimens is not an exercise in artistic ability; it is an exercise in **observation**. Your drawings should therefore show the features that you have observed. See the tips given in the preface to this book.

Your drawings should follow these guidelines:

- Make them as large as the paper you are drawing on will allow.

- Draw them using a **sharp**, preferably HB pencil.

- Draw **sharp outlines** (not 'sketchy' ones).

- Make sure you draw the **same proportions as the specimen** you are drawing. If the anther is, say, three times wider than but only one-sixth as long as the filament, it should be drawn as such. Measure the specimen before you begin and make very faint marks on your paper to guide you. (They can be carefully erased afterwards.)

- Show, if there are any, clear points of structural detail in the specimen; for example, nectar guides on the petal – these points should be shown in the correct place. (If there is a large number of similar such points, only a few need be drawn.)

- Avoid shading. If an area is darker than the rest of the specimen, draw an outline of the area and label it.

- **Rule** label lines in pencil and label in pencil (preferably in capital letters). Label lines should terminate **exactly at the point being labelled**.

- Avoid arrowheads on your labels – they can obscure important features.

- Always give a **magnification** to your drawing. It is a **linear** magnification – that is, calculated by measuring the length or width of the specimen, and the length or width of your drawing measured across the same structural feature.

$$\frac{\text{Length of drawing}}{\text{Length of specimen}} = \text{Magnification (e.g. } \times 4.5)$$

TIP

1 When drawing from a photograph, there is sometimes a magnification given for the photograph. In such cases, you must multiply the magnification of your drawing by this figure.

2 It would not normally be the case that your drawing or measurements would be accurate enough to give a magnification to more than one decimal place.

3 Do not 'round off' too much. Example: ×4.6 is not ×5.

PRACTICAL

Seeing the difference in pollen grains between wind- and insect-pollinated flowers

Using two clean glass slides, gently tap a mature anther from the insect-pollinated flower on one slide, and gently tap the entire wind-pollinated flower on the other slide. View both under the microscope (low power will probably be sufficient).

Compare the appearance of the pollen grains from the two different types of flower.

So long as the pollen grain has arrived on the stigma of the correct species of plant, the sugary solution on the stigma forms a medium in which the pollen grain will germinate.

Events that occur after pollination

So long as the pollen grain has arrived on the stigma of the correct species of plant, the sugary solution on the stigma forms a medium in which the pollen grain will **germinate**.

Germination of the pollen grain involves the growth of a **pollen tube**, which releases **enzymes** at its tip in order to **digest** the cells of the style beneath. In this way, the cells of the style are removed to allow the pollen tube to **grow** down the **style** towards the **ovary.**

> **TIP**
> The style is not a hollow tube down which the pollen tube grows, or falls. It is a solid cellular structure that has to be digested by enzymes from the pollen tube.

Fertilisation in flowering plants

On arrival at the ovary, the end of the pollen tube enters an **ovule** through a small hole called the **micropyle**. Inside the ovule is the **embryo sac**, which contains the female **gamete** within which is the female **nucleus**.

The end of the pollen tube then bursts to release the **male** gamete, which has travelled down the pollen tube from the pollen grain (Figure 16.6). Fertilisation occurs as the male nucleus (carried inside the pollen grain), delivered through the pollen tube, fuses (joins) with the female nucleus in the ovule.

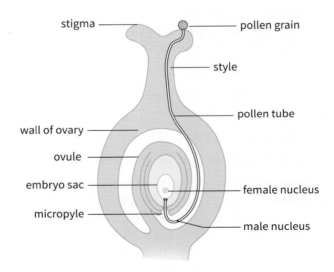

Figure 16.6 Fertilisation in a flowering plant

Worked example

a   In an insect-pollinated flower, explain the importance of nectar and pollen.

b   The diagram show three pollen grains, A, B and C, on part of flower. Describe and explain the similarities and differences between them.

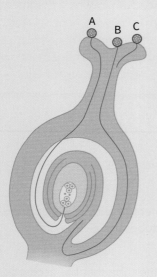

Figure 16.7

Answer

a   The question asks for 'the importance'. It thus requires you to consider the importance of nectar to an insect as well as to a flower. Nectar contains sugar, which is energy-giving food for insects. They are thus attracted to the flower since it is a food source for them.

Pollen contains the male gametes and must be carried from the male part (anther) to the female part (stigma) of a flower in order that fertilisation, and thus sexual reproduction, can occur.

b   The question begins with an exercise in observation. The similarities shared by all three pollen grains is that they have been brought form another flower (i.e. pollination has occurred in all three cases) and all three have grown pollen tubes (i.e. they have germinated).

The differences are that the pollen tube from pollen grain A – probably the first to arrive, has grown through the style, entered the ovary and reached the ovule of the flower. It has entered through the micropyle and fertilisation is about to take place. The pollen tube from pollen grain B has, like A, grown down through the style and reached the ovary. It has remained within the ovary wall and appears unlikely to be able to enter the ovule, whereas pollen grain A has only just germinated and its pollen tube is only about to enter the style. Only one ovule is shown in the ovary, thus it is unlikely that the male nuclei within pollen grains B and C will be involved in fertilisation.

Note that technical terms have been used wherever possible and, notice too, there is a reference to the pollen tube *growing* down the style – not simply passing down or through it.

Progress check 16.3

1   What process indicates the moment of pollination in a flowering plant?

   A   the entry of the pollen tube into the ovule

   B   the germination of a pollen tube

   C   the landing of a pollen grain on a stigma

   D   the release of pollen by a flower

2   A student made a drawing of a leaf. Her drawing was 46 mm long. When she measured the specimen, she found that it was 77 mm long. What was the magnification of her drawing?

Germination of seeds

When suitable environmental conditions are available, the seed will germinate. The conditions necessary for seed germination are:

1 **Water** (to activate the enzymes).

2 **Oxygen** (to allow for the release of a great deal of energy from respiration to fuel the greatly increased growth rate).

3 A **suitable temperature** (enzymes operate efficiently only if the temperature is suitable for them).

(Light is not necessary, except in a very few cases.)

TIP

Be sure to use the three stated terms: water, oxygen and temperature. 'Moisture' is not the same as 'water'. There is moisture in the air, but very rarely enough to allow seeds to germinate. 'Air' contains many gases other than oxygen – none of which are likely to have a positive effect on germination. 'Warmth' is a comparative term. What may be warm enough for some seeds to germinate may not be so for others, thus each seed needs its own suitable temperature for germination.

PRACTICAL

The need for water, oxygen and the right temperature for germination can be demonstrated experimentally as follows:

Apparatus: 4 test-tubes
Cotton wool
Rubber bung
Ignition tube
Alkaline pyrogallol
Cotton
Dry seeds (e.g. cress)

Method:

Set up the experiment as shown in Figure 16.8 and leave the test-tubes for 2–4 days.

Results:

Only the seeds in the tube that has access to all the three conditions (tube A) will germinate. Thus a suitable temperature, water and oxygen are all necessary for germination.

Conclusion:

A **suitable temperature**, **water** and **oxygen** are all necessary for germination.

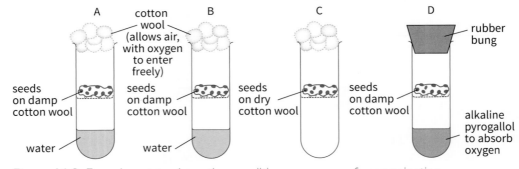

Figure 16.8 Experiment to show the conditions necessary for germination

A The control. Seeds left in air at room temperature (a suitable temperature for germination).

B Seeds left in the refrigerator (4 °C) – not a suitable temperature for germination.

C Seeds left in air at room temperature.

D Seeds left at room temperature.

Chapter summary

- You have learnt the difference between asexual and sexual reproduction.

- You have learnt the structural features of flowers.

- You have learnt that pollination precedes sexual reproduction in flowering plants and that pollination is a process carried out either by wind or by insects.

- You have learnt how flowers have adaptations depending on which type of pollination they undergo.

- You know how fertilisation occurs in flowers and the conditions necessary for seed germination.

Exam-style questions

1   Figure 16.9 shows the structure of a flower.

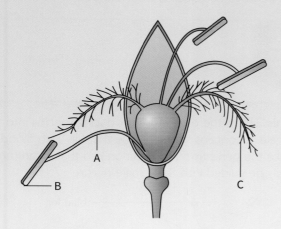

Figure 16.9

a   With reference to features shown in Figure 16.9, state how pollination is most likely to occur in this flower. Your answer should include an identification of structures A, B and C. [4]

b   State three features not shown in Figure 16.9 that a flower with this type of pollination wold be expected to possess. [3]

[Total 7]

2   Explain why a species of plant that reproduces only by cross-pollination is likely to evolve at a faster rate than one that uses only self-pollination. (Chapter 19 will provide some help with this question.) [6]

3   Explain the differences between the pollen grains produced by a wind-pollinated plant and those produced by an insect-pollinated plant. [10]

Reproduction in humans

17.01 Human reproduction

Although a baby is born with a full set of reproductive organs, they are not functional during the first 12 (or so) years of life. Under the influence of **hormones** from the **pituitary** gland, the organs become active at the time known as **puberty**.

The male reproductive system

The male nuclei that are involved in the process of sexual reproduction in humans are located within male **gametes** (sex cells) called **sperms** (an abbreviation for 'spermatozoa'). The male reproductive system is designed to manufacture sperms and to deliver them to the place where they will be able to fuse with a female nucleus.

Testes (singular – testis): These are the **gonads** of the male – that is, they are the organs that produce the gametes. In this case, it is the sperms. Testes are made of millions of tiny coiled tubes. The cells forming the walls of these tubes are constantly dividing to produce up to 100 million sperms per day. The testes work more efficiently at just below body temperature, thus testes are held outside the body in the **scrotum** (or scrotal sac).

Sperm ducts: These are tubes that carry the sperms away from the testes. They join with one another and with the tube bringing urine, at a position just under the bladder.

Prostate gland: This is a gland, about the size of a golf ball, which surrounds the junction between the sperm

ducts and the tube from the bladder. It manufactures, and adds to the sperms, a nutrient fluid (**seminal fluid**) in which the sperms are able to swim.

Sperms + Seminal fluid = Semen

Urethra: This tube carries both urine and semen along the **penis** to be released from the body.

Penis: This is the organ for introducing sperms into the female. It contains spongy tissue that fills with blood to make the penis firm (an 'erection') so that it can more easily be guided into the female (Figure 17.1).

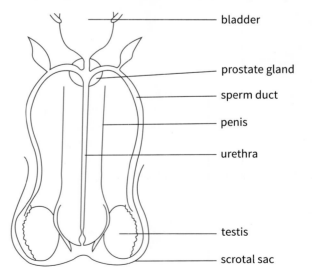

- bladder
- prostate gland
- sperm duct
- penis
- urethra
- testis
- scrotal sac

Figure 17.1 The human male reproductive organs

The female reproductive system

The female nuclei that are involved in the process of sexual reproduction are located in the female gametes called **egg cells** (or **ova** – singular, ovum). The female reproductive system has the function, not only of making ova and ensuring that they are fertilised by the male gametes, but also then to protect and nourish the embryo until it is born (see Figure 17.2).

Ovaries: These are the female **gonads**, making and releasing the female **gametes** (eggs or 'ova'). There are two, each a little smaller than a ping-pong ball, lying in the lower abdomen, Each ovary releases one ovum every 8 weeks, alternately (thus the female releases one ovum every 4 weeks, from alternate ovaries).

Oviducts: These are the tubes which carry the eggs away from the ovaries. They are lined with cilia (see 'trachea'), which help to move the eggs (together with a little muscular assistance) gently along. If **fertilisation** is to occur, it does so about **one-third** of the way along the oviduct.

Uterus: This is a pear-shaped organ lying behind and slightly above the bladder. Its walls contain **involuntary muscle** (i.e. muscle that cannot be consciously controlled). It is where the embryo develops during **pregnancy**.

Cervix: The 'neck' of the uterus, where the uterus joins the vagina. It supplies mucus to the vagina.

Vagina: This is the part of the female system which receives the penis during **copulation**. It is muscular and stretchable (it forms part of the birth canal) and it connects the cervix with the slit-like **vulva** opening to the outside.

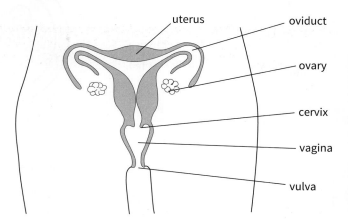

Figure 17.2 The human female reproductive organs

Once sperms have entered the female reproductive system, they swim by lashing their tails (or **flagella** – singular flagellum) through the uterus and up into the oviducts. An egg cell (ovum) is moved along by the cilia that line the oviduct. The journey towards the sperm may take a few days during which time, the ovum relies on the energy stores in its cytoplasm. It is protected by a jelly coating. If one of the fastest-swimming sperms meets an egg cell, it will enter the egg cell, with the aid of enzymes from the acrosome in its head portion. The nucleus of the sperm (the male gamete) will fuse (join) with the nucleus of the female gamete (the egg cell) during the process known as **fertilisation**.

Table 17.1 and Figure 17.3 give information about the male and female gametes.

Male gametes (sperms)	Female gametes (ova)
Released in millions	Released one at a time
Able to move (are motile)	Unable to move on their own
Very small (0.05 mm – of which around 80% is tail)	Comparatively large (0.1 mm in diameter)
Very little cytoplasm	A lot of yolky cytoplasm
Nucleus contains either an X or Y chromosome	Nucleus always contains only an X chromosome

Table 17.1

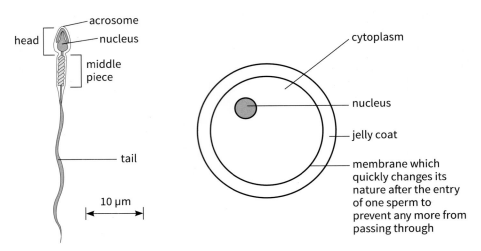

Figure 17.3 A human sperm and a human ovum

The adaptive features of a sperm are:

- its ability to swim by lashing its **flagellum** (the correct name for its tail)
- the large number of **mitochondria** in the middle piece that supply the energy to propel the flagellum
- the presence in the **acrosome** of enzymes capable of digesting the membrane around the egg cell (ovum) and the jelly coat that protects it.

The adaptive features of the egg cell (ovum) are:

- the **jelly coat** and **membrane** that act as protection to the ovum as it passes down the oviduct and then a barrier to the entry of other sperms after fertilisation
- **energy** is stored within the very small amounts of **fat** and **protein** in the cells of the cytoplasm. There is just enough to nourish the cells from the time they are released from the ovary until they arrive, after fertilisation, into the uterus where the zygote absorbs nutrients secreted by the spongy uterus lining.

Progress check 17.2

1 Which part of a sperm contains a large number of mitochondria?

 A acrosome

 B middle piece

 C nucleus

 D tail

2 Why are there so many mitochondria in this part?

Fertile and infertile phases of the menstrual cycle

When there is no ovum in the oviducts, fertilisation cannot occur. It is unlikely to occur if the ovum is not in the correct position in the oviduct, though this is affected by the fact that sperms can live in the oviduct for a few days, allowing the ovum chance to arrive. A woman's **most fertile period** is therefore from a **few days before ovulation** (allowing for the possible survival of sperms in the oviduct) to a **few days after ovulation**. Outside this time, she is less likely to become pregnant.

Human fertilisation

During sexual intercourse or **copulation**, the erect penis is inserted into the vagina. Sensitive cells near the end (glans) of the penis are the receptors for a **reflex** action leading to the release (**ejaculation**) of semen. Up to 300 million sperms are deposited near the cervix of the female, from where they swim through the uterus and up the **oviducts**. If they meet an egg cell around **one-third** of its way from the ovary, **one** of the sperms will fuse with the ovum. When their nuclei join, a zygote is formed. This is the moment of **fertilisation**.

Development of the zygote

The zygote, a single cell formed from equal nuclear contributions from both parents, now begins to divide (by **mitosis**), eventually to form an **embryo**, which, in the first instance, is a **hollow ball of cells**. On arrival at the uterus, whose walls are covered with a spongy lining containing blood capillaries, the embryo sinks into the lining and becomes embedded in it. This process is called **implantation**, after which the embryo is referred to as a fetus. The fetus then begins a process of development during which its cells begin to become specialised to perform certain functions. The fetus's organs begin to develop during the (just over) 9 months (the gestation period) that it remains within the uterus.

The functions of the amniotic sac and amniotic fluid

At first the embryo absorbs nourishment secreted by the cells of the uterus, but it soon implants in the spongy lining of the uterus, after which further division of the cells turn the embryo into a **fetus**. The fetus is surrounded by a membrane (the amnion), which forms the **amniotic sac** enclosing the fetus in a water bath (the amniotic fluid). Amniotic fluid is secreted by the amniotic sac.

Functions of the amniotic fluid are:

* to protect the embryo from physical damage – for example, if mother falls over

* to support the embryo, keeping pressure even all round it, allowing organs to develop without restriction

* to allow the fetus some restricted movement.

The nutrition and excretion of the fetus

Both nutrition and excretion are carried out through a special structure called the placenta. This is made up partly of material from the fetus, and partly of material from the spongy lining of the uterus. In the placenta, **blood** in mother's capillaries runs **very close** to blood in the capillaries of the fetus. Food substances and oxygen pass by diffusion from the mother's blood system to that of the fetus, while excretory substances pass in the opposite direction.

> **TIP**
> The **separate bloods do not mix** – they may be different blood groups and the mother's blood pressure would be high enough to cause damage to the fetal system.

Diffusion of various substances takes place between the two blood systems (Table 17.2).

Diffusing from mother to fetus	Diffusing from fetus to mother

dissolved nutrients:
  glucose
  amino acids
  ions
  vitamins
  water
	nitrogenous waste:
	  urea

dissolved gas:
  oxygen
	dissolved gas:
	  carbon dioxide

Table 17.2

While the placenta allows the passage of necessary substances (including some antibodies), it acts as a barrier to some, but not all, potentially harmful ones.

Most toxins and disease-causing (pathogenic) organisms do not pass from mother to embryo. Those that do pass include nicotine and carbon monoxide (from cigarettes) and the virus that causes AIDS (the human immuno-deficiency virus – HIV) and the rubella virus that causes German measles. The disease German measles can interfere with the normal development of a fetus inside a pregnant mother who catches the infection. These are all carried to the fetus by the **umbilical cord**, inside which run **fetal** blood vessels. The **umbilical vein** brings substances **to** the fetus, an **umbilical artery** carries substances **from** the fetus.

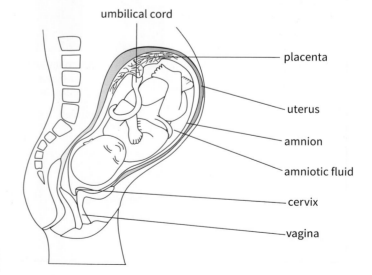

Figure 17.4 A fetus inside a pregnant woman

During early development of the fetus, the hollow ball of cells gradually increases in complexity as the organs of the body develop but during the later stages of pregnancy, those organs and thus the fetus in general, gradually increase in size.

The dietary needs of a pregnant woman

Since an embryo's development depends on the food eaten by its mother, a pregnant woman must adjust her diet accordingly. She should ensure that the levels of the following constituents are higher than in her normal intake:

- **Protein** for the manufacture of embryonic tissues.

- **Carbohydrate** for addition respiration in embryonic tissues.

These should be raised to the approximate levels required by a very active woman who is not pregnant.

- **Vitamin C** for making proteins in the embryo

- **Vitamin D** and calcium for making bones and teeth of the embryo

- **Iron** for making the embryo's blood.

A pregnant woman should also pay attention to her **general life-style**. The following are all recommended for a successful pregnancy and for the birth of a healthy baby:

- Take regular exercise.

- Avoid smoking – carbon monoxide in cigarette smoke decreases the amount of oxygen that can be carried in the baby's blood, retarding its development.

- Avoid alcohol – it can result in the baby being born with developmental problems, including hearing, liver, kidney and heart defects.

- Attend regular medical checks.

Worked example

a   Explain why a woman should avoid smoking during pregnancy.

b   Explain the information provided by the graph, which shows the requirement for protein in the diet of three women.

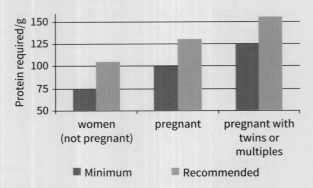

Figure 17.5

Answer

a   You may need to check on the harmful constituents of cigarette smoke but the important one here is carbon monoxide. As a

fetus develops, much cell division will occur and this requires a great deal of energy. Respiration provides energy, and respiration requires oxygen. The fetus's oxygen is supplied by its mother, but if the mother smokes, then her hemoglobin, having absorbed carbon monoxide, is unable to carry the normal amount of oxygen. Reduced oxygen to the fetus results in impaired cell division and therefore the growth of the fetus is restricted. The baby is likely to be born underweight and not fully developed. Toxins in cigarette smoke may also pass to the baby adversely affecting its growth and development.

b   The overall conclusion from the graph is that pregnancy requires additional protein, though the recommended amount of protein for a non-pregnant woman is adequate if maintained during her pregnancy (unless she is expecting twins). The explanation of this is that protein is required for growth. The mother is likely to have finished growing but the additional protein is required for supplying the fetus with enough protein to grow and develop well throughout pregnancy.

Labour and birth

Just over 9 months after fertilisation, and under the control of hormones from the brain, the **muscular walls of the uterus begin to contract** rhythmically. This is the beginning of **labour**. At first the time between each wave of contractions will usually be in excess of ten minutes, but the contractions gradually become more frequent and longer lasting. The baby, which should now be lying with its head pointing downwards, begins to be forced towards the cervix. The **cervix widens (dilates)** to allow the baby to pass and the **amniotic sac bursts** allowing the amniotic fluid to escape. The baby begins to be pushed **through the vagina**, then, once the head of the baby has emerged through the vulva, the rest of the baby follows quite quickly.

Immediately the baby is born, the **umbilical cord is cut and sealed (tied)** and remains attached to the placenta, which is still attached to the uterus. This will later be expelled and is known as the **afterbirth**.

The value of breast feeding

After the birth of the baby, **milk** from the **mammary glands** supplies the ideal food for the first months of development, since it:

- contains all the necessary **constituents**, in the **correct proportions**

- is at the **correct temperature**
- contains some **antibodies** that protect the baby against disease
- develops the **mother/baby bond**
- is **cheap**
- is readily **available**.

Some babies have suffered reactions to the substitute formula milk powders that are used in bottle feeding, and the bottle, milk powder or water used in preparing the milk may contain pathogenic bacteria causing the baby to suffer from infection.

17.02 Sex hormones in humans

Puberty is the stage in a person's life when they change from being a child to being an adult capable of reproduction. The changes in their bodies that occur at this time are the result of hormones that their gonads (testes in males, ovaries in females) begin to secrete (Table 17.3).

These changes are known as secondary sexual characteristics.

Name of hormone	Where produced	Effect of hormone
Males: testosterone	produced by the testes	• growth of hair on face and body (under arms and above genitals) • larger larynx (voice box) and deeper voice • larger penis and testes • the production of sperms • development of stronger muscles
Females: oestrogen	produced by the ovaries	• development of mammary glands (breasts) • widening of hips • growth of hair under arms and above genitals • release of eggs and start of the menstrual cycle • fat deposited under skin giving body a more rounded shape

Table 17.3

Menstrual cycle

Once a female reaches puberty, she will start to **ovulate** (release eggs or ova from her ovaries). Ovulation is one stage in her **menstrual** or monthly cycle. This cycle is associated with the production of a **spongy lining** to the walls of the uterus in which the fertilised ovum will develop. The spongy lining contains many blood capillaries.

Ovulation occurs and the ovum begins its journey down the oviduct. By the time it reaches the uterus, the lining of the uterus will have developed ready to receive the ovum if it has been fertilised. But, if unfertilised, the ovum carries on through the uterus and vagina and out of the vulva. During this process, development of further ova from the ovaries is put 'on hold'.

The spongy lining, now redundant, peels away from the uterus wall, damaging the blood capillaries, and is passed out of the vagina and vulva together with blood. This is **menstruation** or the monthly period, which lasts for a few days and occurs about 2 weeks after ovulation (Figure 17.6).

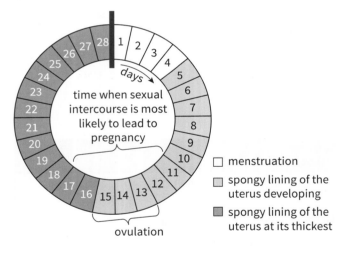

Figure 17.6 The human female menstrual cycle

Once the uterus wall has recovered, it begins to rebuild its spongy lining (under the influence of a hormone from the ovary). Meanwhile, a new ovum is maturing in the ovary (under the influence of a hormone from the pituitary) and when mature, the ovum is released (ovulation), around 2 weeks after menstruation.

This cycle continues until an ovum is fertilised, when the female is then **pregnant** or until the female reaches the menopause when, at around 50 years of age, she stops ovulating and thus can no longer become pregnant.

Hormonal control of the menstrual cycle

The control of the menstrual cycle involves several hormones, some produced by the **pituitary gland** under the brain and some by the **ovaries**.

The four significant hormones and their functions are shown in Table 17.4:

Hormone	Where produced	Function
FSH	Pituitary gland	Promotes the ripening and release of eggs from the ovary
LH	Pituitary gland	Works with FSH to help ripen and release eggs, also prompts the ovary to release progesterone and oestrogen
Oestrogen	Ovary	Helps to create spongy lining to uterus wall after menstruation and before ovulation
Progesterone	Ovary	Secreted after ovulation to maintain the spongy lining to the uterus throughout pregnancy. Stops ovulation by inhibiting the release of FSH. The ovary ceases to produce progesterone if fertilisation does not occur.

Table 17.4

The cells in the ovary that surround and nourish each egg before it is released (**follicle cells**), are the ones that become a temporary ductless gland after its release. This temporary ductless (or 'endocrine') gland is called the **yellow body** and, under stimulation from LH, releases progesterone. If fertilisation does not occur, the yellow body breaks down after about 2 weeks, thus progesterone is no longer produced. The production of **FSH** is thus no longer inhibited, so the cycle begins again.

17.03 Methods of birth control in humans

The world population has risen alarmingly over the last few decades. It is already difficult to supply enough food to all areas of the world in adequate quantities. A solution lies in birth control. The main methods are discussed here.

Natural method

The 'natural' method (rhythm method) in which sexual intercourse (copulation) is limited only to those times in the menstrual cycle when fertilisation is less likely.

Abstinence from sexual intercourse is also regarded as a 'natural' method of birth control.

Not all healthy and perfectly happy women have predictable and regular periods — particularly when teenagers or when approaching the menopause — and the most accurate prediction can only be that ovulation usually occurs somewhere between 12 and 16 days before the start of menstruation. It is therefore not a reliable method of avoiding pregnancy. (Severe anxiety and malnutrition can suppress the menstrual cycle completely. Varying degrees of anxiety and dietary deficiency can lead to an erratic menstrual pattern.)

Monitoring body temperature over a period of time can inform a woman when she is likely to ovulate. Two or three days after ovulation, the body temperature rises by up to 0.5 °C. If this rise is recorded, the next ovulation can be predicted in 25–26 days' time.

Cervical mucus can also help tell about when she is fertile. The mucus that is released from the cervix of the uterus changes its nature during the menstrual cycle. The mucus is clear and slippery towards the time when the woman is at her most fertile. When she is least fertile, it is thick and cloudy.

Withdrawal of the penis from the vagina just before ejaculation is not a reliable method of natural birth control. A few sperms may be released before full ejaculation, and one of those may fertilise an ovum. Ejaculation may occur before withdrawal, despite contrary intentions.

Chemical methods

1   Spermicides: These are the chemicals that kill sperms. They are put into the vagina of the female before intercourse. They are not very effective on their own and are often used in conjunction with a mechanical method.

2   Contraceptive pill: This must be taken regularly by the woman. The pill contains a hormone that prevents ovulation, therefore, no ova are present to be fertilised. It is effective if the routine is followed carefully.

3   A hormone injection received every 12 weeks. One type stops the ovaries from releasing egg cells.

4   A contraceptive patch that is stuck onto the arm and delivers hormones similar to those in the pill. A new patch is applied every week for three weeks, followed by one 'patch-free' week.

5   A contraceptive implant is a small, thin, flexible, plastic tube that is implanted under the skin of the arm. It slowly releases a progesterone-like hormone that makes the mucus released by the cervix thick and difficult for sperms to swim through. It also stops ovulation and makes the spongy lining of the uterus too thin to support a developing fetus. It can last for up to 3 years and is very reliable.

Hormonal regulation of fertility has particular advantages.

• It is easy to use.

• It decreases blood flow during menstruation — thus decreasing iron loss in the blood.

• Pills and injections are reliable forms of contraception.

• Pills and injections are safe and help to guard against a number of diseases of the reproductive system.

• Hormones (FSH and LH) can also be used to stimulate egg development and ovulation and help women who would not otherwise become pregnant to have children.

Mechanical methods

1   Some form of barrier is put between the sperms and the ova. Barrier methods are popular and quite effective. The barrier may take the form of:

a)   a condom, a sheath placed over the penis before intercourse to catch the sperms when ejaculated

b)   a femidom, which lines the vagina of the female with the same result

c) a **diaphragm**, which fits over the cervix of the uterus, preventing the entry of sperms. It is usually used with a spermicide.

2 There is no barrier between the sperm and the ovum. An **inter-uterine device or system** (IUD or IUS) is fitted inside the uterus. It does not stop fertilisation, but it prevents implantation of the **blastocyst**. It is effective.

Surgical methods

1 In the **male** – cutting the sperm ducts (the operation is called **vasectomy**). It is effective but rarely reversible.

2 In the **female** – tying the oviducts to prevent the passage of ova. It is effective and usually reversible.

> **TIP**
>
> Abortion, that is, the surgical removal from the uterus of an unborn fetus, is **not** considered to be a method of birth control.

Progress check 17.3

1 Why is it more common for women than men to require iron supplements in their diet?

2 Why is withdrawal of the penis before ejaculation not a reliable method of birth control?

Artificial insemination

This is the introduction, into the female's vagina or uterus, of sperms from a male using a technique other than natural sexual intercourse. It may be used when the male partner is infertile or incapable of normal intercourse (e.g. the inability to achieve and sustain an erection) or when the female has no male partner and donor sperms are then used.

Fertility drugs and the social implications of using them: *in vitro* fertilisation

If a woman has problems with her oviducts that prevent the egg cell from travelling along them, or if the male's sperm count is low or perhaps only a few of his sperms are suitable for the process of fertilisation, an egg cell may be taken from the woman at the appropriate time in her menstrual cycle and exposed under laboratory conditions to the man's sperms. Only when successful fertilisation has occurred is the egg cell then returned to the woman's uterus to continue its development. This is known as *in vitro* fertilisation (IVF) and allows couples who would otherwise be childless to have children.

Social implications

People who would otherwise be childless are thus allowed to have a much wished-for family, but there are those who believe that methods other than natural ones should not be used – that is, it should not be considered to be a right to have children. Also, fertility treatment is expensive. It may be considered to be available only to the wealthy. It can often give rise to multiple births, which may stretch the resources available for looking after the children.

Some believe that it is wrong to interfere with nature and that nature should be allowed to 'run its course'. (Also, see Chapter 8, under 'Pollution', for the effects of contraceptive hormones on water courses.)

17.04 Sexually transmitted infections

During the act of sexual intercourse, the bodies of the partners are brought into close contact during which **pathogens** can easily pass via body fluids from one person to another. Sexually transmitted infections (STIs) include **HIV/AIDS**. AIDS is caused by HIV (the human immuno-deficiency virus). AIDS is a condition where the body's ability to resist infections is progressively reduced until the situation is reached where a relatively minor disease can cause serious illness – even death.

The definition of a **sexually transmitted infection** is an **infection that is transmitted via body fluids through sexual contact**.

AIDS

AIDS is a collection of symptoms that are caused by an infection with the virus known as HIV. HIV is found in **body fluids** such as **blood** and **semen**. It is thus transmitted from host to host when drug users **share unsterilised needles** to inject themselves, since there is usually a little blood from the previous user which is injected along with the drug into the second (and subsequent) user.

A very small number of people who are infected with HIV may not suffer from AIDS without treatment and, with appropriate drug treatment, others may avoid suffering from the more severe aspects of aids for many years

It is transmitted in **semen** from one partner to the blood of another if there is any tearing of tissues during intercourse.

It is transmitted from an infected mother's blood to her baby's blood during the birth process.

It may by accidentally transmitted in untreated blood during **blood transfusion**.

Although some drugs can slow the progress of the disease, there is, as yet, no cure for it. It can, however, be controlled by:

1 Educating the public about how it is spread and what precautions can be taken.

2 Never sharing needles.

3 Avoiding sex with prostitutes, since they are often carriers of the disease. Stay with one STD-free partner.

4 Always using a condom if in doubt, or other barrier method of contraception that prevents direct contact between the body fluids of the two partners.

5 Treating all blood and blood products used in transfusions to destroy the AIDS virus.

All the controls are also effective, to a greater or lesser degree, against other forms of STI.

How HIV/AIDS affects the immune system

When a pathogen invades the body, its proteins act as antigens against which lymphocytes in the blood produce specifically engineered antibodies to help destroy the pathogen.

The AIDS virus (HIV), however, targets a particular type of lymphocyte (known as a **helper T-lymphocyte**). It uses the helper T-cell for its own reproduction, killing the cell. More viruses are released to attack further T-cells. The job of helper T-lymphocytes is to recognise foreign proteins and produce substances that activate other lymphocytes to produce antibodies. With the decline in number of helper T-cells, the body's resistance to many infections decreases. Eventually the stage is reached when a relatively minor ('opportunistic') infection may develop into a serious, possibly fatal condition, as insufficient antibodies can be made to resist and control it.

Chapter summary

■ You have learnt the names and functions of the parts of the male and female reproductive systems.

■ You have learnt about fertilisation and the development and birth of the fetus.

■ You have learnt about human sex hormones.

■ Students following the supplementary course have learnt about the sex hormones controlling the menstrual cycle.

■ You have learnt about methods of birth control and about sexually transmitted infections.

Exam-style questions

1. a Humans use a variety of methods of birth control. Explain how a surgical method of birth control prevents a male from fathering children. [4]

   b Explain how a condom can be both a method of birth control and a way of reducing the spread of a **named** sexually transmitted infection. [6]

   [Total 10]

2. a i) State a male secondary sexual characteristic.

      ii) Name the hormone responsible for the appearance of this characteristic.

      iii) Name the organ that produces this hormone in the male. [3]

   b Figure 17.7 shows the changes in concentration of four hormones in the blood of a woman during one menstrual cycle.

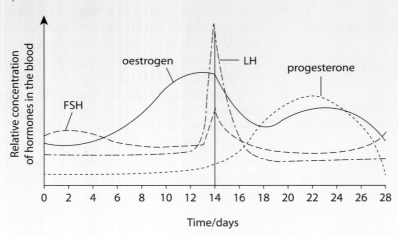

Figure 17.7

   i) Describe the changes of concentration of progesterone during one menstrual cycle. [3]

   ii) Explain these changes in relation to the function of this hormone. [5]

   [Total 11]

3. State the origin of excretory products in a fetus and describe how they pass from where they are produced to the environment in which the mother and fetus are living. [10]

Inheritance

18.01 Inheritance

The features that are characteristic of all individual organisms have been passed on to (inherited by) them from their parent(s). These characteristics are controlled by the genes on the chromosomes of their nuclei.

Inheritance is defined as **the transmission of genetic information from generation to generation**.

18.02 Chromosomes and making proteins

A **chromosome** is a **thread-like structure of DNA (or deoxyribose nucleic acid) carrying genetic information in the form of genes**.

The unit of inheritance

A **gene** is defined as **a length of DNA that is the unit of heredity (inheritance) and codes for (is responsible for the production of) a specific protein**.

A gene is a part of a chromosome and can be copied and passed on from parent to offspring via chromosomes in the nuclei of the parents' gametes.

For the purposes of understanding simple inheritance, it is convenient to imagine a chromosome as a string of beads, like that shown in Figure 18.1. Each bead represents a gene.

Figure 18.1 Representation of a chromosome

Most features (characters) found in an individual are the result of the interaction between one gene inherited from the father and one gene, occupying exactly the same position on the exactly matching (or homologous) chromosome, from the mother. Such a pair of matching genes are described as **alleles** (Figure 18.2).

Although the correct term is 'character' it is commonly referred to as a 'characteristic'.

An **allele** is defined as a **version of a gene**.

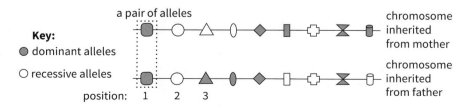

Figure 18.2 Representation of alleles on homologous chromosomes

Chromosomes are found in **nuclei** but in the higher organisms, including the flowering plant and the human being, chromosomes may be present in matching ('**homologous**') pairs (one of each pair having been inherited from the male parent, the other from the female parent) in which case, the nucleus is described as **diploid**. Alternatively, the chromosomes may be present in unmatched single strands, in which case, the nucleus is described as **haploid**.

Thus a **haploid nucleus** is defined as a **nucleus containing a single set of unpaired chromosomes (e.g. a sperm and an egg).**

A **diploid nucleus** is a **nucleus containing two sets of chromosomes (e.g. in body cells).**

TIP

It might be easier to remember the meaning of these two term if you think of 'ha' as in 'half' and 'di', which means 'two' as in 'dichotomous' and carbon 'dioxide'.

The inheritance of sex

Whether a child is born male or female is determined at the moment of fertilisation.

Of the 23 pairs of chromosomes in a human nucleus, one pair (the 23rd pair) is known as the sex chromosomes. In the female, the sex chromosomes are identical and are called X chromosomes.

In the male, they do not exactly match one another. One of them is an X chromosome, exactly like those in the female, but the other is a (shorter) Y chromosome. Thus, the sex chromosomes are:

XX for a female

XY for a male

The gametes contain 23 single chromosomes and thus only one of the two sex chromosomes that exist in normal body cells.

In females, all gametes contain an X chromosome (she has no other type to give to her gametes).

In males, 50% of the gametes contain an X chromosome and 50% contain a Y chromosome.

Thus there is an exactly equal chance of the X chromosomes in the ovum fusing with a X-carrying sperm to produce a daughter, or a Y-carrying sperm to produce a son.

Parents

Father × Mother

		Father	Mother
Sex chromosomes:	in body cells	XY	XX
	in gametes	X or Y	X (only)

At fertilisation chromosomes in male gametes	→	X	Y
	X	XX	XY

↑

chromosomes in female gametes

Chromosomes in offspring probability	female (XX) 50%	male (XY) 50%

Progress check 18.1

1   When is the sex of a baby determined?

2   Which contain Y chromosomes in humans?

   A   all egg cells

   B   all sperm cells

   C   half of all egg cells

   D   half of all sperm cells

DNA and protein production

Since a gene is a length of DNA, as we have seen in Chapter 4, it follows that a gene will be made up of a sequence of paired bases. This sequence is responsible for the construction of proteins in the **ribosomes** found in the cytoplasm. The sequence of bases on any particular chromosomes fixes the order in which amino acids become linked together in a ribosome to form a specific protein.

Some of the proteins so constructed will be **enzymes** (as all enzymes are proteins). These enzymes will control all chemical reactions taking place in the cell, thus DNA controls all cell functions.

Other proteins manufactured for purposes other than growth include **antibodies** (see Chapter 9, 'White blood cells' and Chapter 10, 'How antibodies work to kill pathogens') and **receptors** for neurotransmitters at synapses (see Chapter 13 'Synapses').

How DNA controls protein production

- DNA is found in the nucleus, but protein production occurs in the ribosomes found in the cytoplasm.

- There must, therefore, be a way for the information contained in the genes of a chromosome to be sent to the ribosomes

- This task is performed by **messenger RNA** (normally referred to as **mRNA**).

DNA is a molecule of two twisted strands linked by pairs of bases. RNA is similar to just one of those strands, with a row of single bases attached along it. The difference between a strand of RNA and one of the two strands in a molecule of DNA is (i) the type of sugar found within each strand and (ii) the base 'T' occurs in DNA, but not in RNA, and the base 'U' appears in RNA but not DNA.

To make a protein molecule, the following sequence of events occurs:

- In the nucleus, along a section of a chromosome (gene) the bases separate.

- A molecule of mRNA is built up with the sequence of bases along its length fixed by the sequence of exposed bases on one of the strands of the DNA.

- The mRNA strand then leaves the nucleus, carrying its copy of the base sequence (see Figure 18.4) to a ribosome in the cytoplasm.

- On arrival at a ribosome, the strand of mRNA passes through the ribosome.

- The ribosome 'reads' the information on the mRNA to make a protein.

- Each successive run of three bases in sequence will decide which amino acid will be added to the growing protein molecule. (The run of three bases was fixed by three successive bases (a triplet) on the DNA molecule in the nucleus.)

- When the mRNA has passed through the ribosome, a new protein molecule has been built.

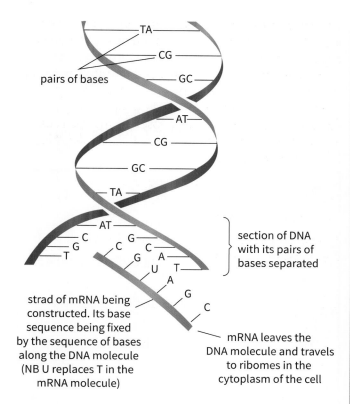

pairs of bases

section of DNA with its pairs of bases separated

strad of mRNA being constructed. Its base sequence being fixed by the sequence of bases along the DNA molecule (NB U replaces T in the mRNA molecule)

mRNA leaves the DNA molecule and travels to ribomes in the cytoplasm of the cell

Figure 18.3 Constructing a strand of mRNA

The chromosomes in a nucleus contain many thousands of genes, each gene is 'switched on' only when there is a need to make the particular protein for which it codes. Those cells that 'specialise' in the production of a particular protein may only rarely make some of the others, even though their nucleus possesses the gene for doing so.

The arrangement of chromosomes in a nucleus

In the flowering plant and the human being, chromosomes may either be present in matching pairs (one of each pair having been inherited from the male parent, the other from the female parent), in which case the nucleus is described as **diploid**, or the chromosomes may be present in unmatched single strands, in which case the nucleus is described as **haploid**.

Thus a **haploid nucleus** is defined as a **nucleus containing a single set of unpaired chromosomes (e.g. in a sperm and in an egg cell, i.e. gametes).**

A **diploid nucleus** is a **nucleus containing two sets of chromosomes (e.g. in body cells).**

Each species has its own set number of chromosomes in each nucleus (its 'chromosome number'). This is a major reason why different species are unable to interbreed – the chromosomes, and thus their genes in their gametes, do not match.

The human chromosome number (i.e. the diploid number) is 46, made up of **23 pairs** (of which the 23rd pair are the **sex** chromosomes).

Progress check 18.2

1   What happens when proteins are made in a cell?

    A   DNA passes from ribosomes into the nucleus

    B   DNA passes from the nucleus to the ribosomes

    C   RNA passes from ribosomes to the nucleus

    D   RNA passes from the nucleus to the ribosomes

2   If the base sequence on a DNA strand is G-A-A-T-C-T, what will be the base sequence on the strand of mRNA for which this sequence codes?

18.03 Cell division

Mitosis

There are a number of occasions when an organism will need to increase the number of cells it possesses. When this happens, it is important that every new cell produced contains the **same genetic information** as all the other cells in the organism's body – that is, the cells produced must be genetically identical not only to the cell that divided to produce it, but also to all other cells in the body. This is achieved by a process of cell division called **mitosis**.

Mitosis is thus defined as **nuclear division giving rise to genetically identical cells**.

There are specific processes that occur in organisms that require the production of more, identical cells. These are:

• Growth

• Repairing damaged tissues

• Replacing cells that wear out

• Producing new organisms by asexual reproduction.

Meiosis

This is the name given to the special type of cell division that produces cells (**gametes**) that have **half** the normal number of chromosomes so that the normal number is restored at fertilisation (when two gametes fuse).

Meiosis is thus a **nuclear division giving rise to cells that are genetically different**.

TIP — Mitosis and meiosis are two words you must learn to spell correctly and you must also be sure of the differences between the two processes.

It has already been stated that genes (and therefore also chromosomes) can be copied. This copying process occurs just before a nucleus begin to divide by mitosis. The newly manufactured chromosomes then separate as the cell divides, ensuring that the two new cells each have the same number of chromosomes and the correct chromosome number for the species.

Therefore, the **chromosome number is maintained by the exact duplication of chromosomes**.

Stem cells

Any new cell has a basic, unspecialised form, but it then has to undergo a process of modification so that it can perform the same functions as other cells in the tissue of which it is to become a part.

It is possible to treat some forms of blood and bone cancer by destroying the cancerous cells and replacing them with unmodified cells from a (matching) donor. The unmodified cells used are called **stem cells** and, as they are unmodified, they have the ability to take on the required form and function in the patient who receives them. Stem cells are usually taken from a donor's bone marrow or fatty tissue. Stem cells are also sometimes 'harvested' from the umbilical cords of a new-born baby.

Stem cells that have a particular ability to form a wide range of tissues are those from an embryo that is up to 5 days old. Their use is, however, controversial as harvesting could result in serious harm to the embryo.

Work is on-going on the use of stem cells to construct replacement organs that might then be implanted.

Meiosis

In organisms that reproduce by **sexual** reproduction, in order that the zygote produced at fertilisation contains the diploid number of chromosomes for that species, the gametes that fuse must contain chromosomes that lie singly (i.e. their nuclei are haploid). Thus, gametes must be produced by a type of division that halves the number of chromosomes. This type of division is called reduction division or **meiosis.**

Meiosis contains a further added effect. During each division process, alleles inherited from the organism's mother and father are randomly exchanged ('shuffled') so that each new chromosome formed contains its own unique combination of alleles. Thus, all gametes are genetically different and it therefore follows that all offspring will be genetically different, leading to an infinite **variation** in the offspring of any two parents.

Meiosis is therefore a reduction division in which the chromosome number is halved from diploid to haploid, giving rise to cells that are genetically different.

Progress check 18.3

1   What refers **only** to the process of mitosis?

   A   Each new cell has a different combination of genes.

   B   It occurs only during gamete formation.

   C   The cells produced are always haploid.

   D   There is an exact duplication of chromosomes.

2   Why are stem cells so valuable when performing tissue transplants?

18.04 Monohybrid inheritance

A character(istic) that shows discontinuous variation (see later) will often exist in only two forms. The alleles that control it also exist in two forms, called dominant and recessive (see Figure 18.2).

An offspring may therefore inherit either:

1   Two dominant alleles (one from each parent) and is then described as homozygous dominant,

2   Two recessive alleles, one from each parent, and is then described as homozygous recessive, or

3   One dominant and one recessive allele and is then described as heterozygous.

These are the three possible genotypes of the individual.

So long as there is at least one dominant allele present in the genotype, then the individual will show the dominant feature in their appearance (or phenotype). Thus the homozygous dominant and heterozygous genotypes will give the same phenotype. The homozygous recessive individual will have the alternative (or contrasting) phenotype (see Figures 18.1 and 18.2).

The appearance of an organism will be affected and modified by the environment in which it lives, as well as the genes it inherits.

The study of inheritance falls within the topic known as *genetics* that contains a number of terms that are specific to this topic.

Definitions

Dominant

Dominant is an allele that is expressed if it is present.

Recessive

Recessive is an allele that is expressed only when there is no dominant allele of the gene present.

Genotype

Genotype is *the* genetic makeup of an organism in terms of the alleles present.

Phenotype

Phenotype is the physical or other observable features of an organism (due to both its genotype and its environment).

Homozygous

Homozygous is having two identical alleles of a particular gene.

Two identical homozygous individuals that breed together will be pure breeding (i.e. also produce similar homozygous offspring).

Heterozygous

Heterozygous is having two different alleles of a particular gene.

A heterozygous individual will therefore not be pure bred.

Progress check 18.4

1   In monohybrid inheritance, how many alleles for a particular character(istic) are inherited from each parent?

2   How many dominant alleles for a particular characteristic does a heterozygous individual possess?

3   What is the correct term for the observable features of an organism?

Organisms inherit alleles for thousands of different contrasting characters. If only one pair of contrasting characters, controlled in the individual by one pair of alleles is being considered, then this is described as monohybrid inheritance.

Example

In genetic diagrams, it is customary to use the same letter, in upper and lower case forms, to represent a pair of alleles. The **upper case** represents the **dominant** allele and the **lower case** the **recessive** allele.

In mice, brown coat colour is dominant over grey coat colour. In an experiment, a homozygous dominant (or 'pure-breeding') brown male mouse mated with a homozygous recessive (also pure-breeding) grey female mouse. All their offspring (i.e. the F_1 or first filial generation) were found to be brown.

The F_1 generation were then allowed to freely interbreed with one-another. It was found that their offspring (the F_2 generation) were brown to grey in a $3:1$ ratio. This can be explained in a genetic diagram, which must be set out as follows:

Let B represent the dominant allele for brown coat colour in mice and b its recessive allele for grey coat colour.

Parents:	male					female	
genotype:	BB		×			bb	
phenotype:	brown					grey	
Gametes:	B	B	× b		b		b

	b	b
B	Bb	Bb
B	Bb	Bb

F_1 generation

This method of working out the genotypes of the offspring is called a Punnett square

genotype: all Bb
phenotype: brown

F_1 selfed   Bb   ×   Bb

(allowed to interbreed)

Gametes: B   b   B   b

	B	b
B	BB	Bb
b	Bb	bb

F_2 generation

possible genotypes:	BB	Bb	Bb	bb
phenotypes:	brown	brown	brown	grey
ratio in a large sample:		3 brown	:	1 grey

The results are given as a statistical ratio in a large sample. The smaller the sample, the less likely it is that the ratios will be exactly as shown.

In humans, where only one offspring is likely to be produced at a time, the probability, often expressed as a percentage, is usually given.

> **TIP**
>
> Make sure that you learn how to set out a genetic diagram, including the identifications of each line of the diagram that appear on the left of it.

Cystic fibrosis

Most people produce normal protein in the mucus found in their lungs. This is due to the possession of at least one dominant allele, *F*. The homozygous recessive person has the genotype *ff*, and their lungs contain some particularly thick and sticky mucus, making gaseous exchange difficult. The person suffers from cystic fibrosis.

If two parents, both heterozygous for this condition (*Ff*) have a child, then the probability of this child having the genotype *ff*, and therefore suffering from cystic fibrosis, is 25%.

A genetic cross producing offspring in a 1:1 ratio

A 1:1 ratio of phenotypes in the offspring is obtained if one of the parents is heterozygous and the other is homozygous recessive. This might be the case with a brown and a grey mouse as shown here:

Parents:

genotypes	Bb	×		bb
phenotypes	brown			grey
Gametes:	B	b	B	b

	B	b
b	Bb	bb
b	Bb	bb

Offspring:

genotypes	Bb	bb
phenotypes	brown	grey
Ratio in a large sample	1:1	

Progress check 18.5

1   Write out a genetic diagram to show the possible offspring from one parent that is heterozygous and the other that is homozygous dominant

2   In an organism whose genotype for a particular character is expressed as *Bb*, what does the '*b*' stand for?

   A   a dominant gene

   B   a heterozygous gene

   C   a homozygous allele

   D   a recessive allele

3   If a heterozygous black-haired mammal breeds with a homozygous white-haired mammal of the same species, what percentage of the offspring will be homozygous dominant?

   A   0

   B   50

   C   75

   D   100

Worked example

In humans, the ability to taste the chemical PTC is the result of possessing the dominant allele T. A man who could taste PTC, and a woman who could not, together produced three children who were all able to taste the chemical. The man concluded that they would thus always produce children who could taste PTC. Explain whether he was correct in making this statement.

Answer

This question combines the requirement for knowledge of monohybrid inheritance with the knowledge that the theoretical results of inheriting features are based on probability.

Both the man and his wife have two alleles controlling their abilities to taste PTC. His wife, being a non-taster, would have the genotype *tt* (i.e. she was homozygous recessive). The man, being a taster, might be either *TT* or *Tt*.

If the man was *TT* (i.e. homozygous dominant) then all his gametes would carry the *T* allele. All his wife's alleles would carry the *t* allele and they would always produce children who had the genotype *Tt* – that is, were tasters. But if he was heterozygous (*Tt*), then half his gametes would carry the *T* allele and half would carry the *t* allele and there would be a 50% chance of producing children who were tasters (*Tt*) and a 50% chance of producing children who were non-tasters (*tt*). The fact that his first three children were all tasters is not a large enough sample to be statistically reliable – his fourth child could be a non-taster. He was thus not correct to make this statement.

(All this information could be supplied in a genetic diagram, but the diagram would require explanation at each stage.)

The test (or back) cross

Of the brown mice in the F_2 generation in the first example given previously, approximately one-third of them will be homozygous dominant (*BB*), and two thirds will be heterozygous (*Bb*) There is no way of telling from their phenotype. Therefore, a test (or 'back') cross is performed.

In a **test cross**, the individual is mated with a homozygous recessive partner. If all the offspring are brown, then the individual must have been homozygous dominant (*BB*); if the offspring are brown:grey in a 1:1 ratio, then the individual must have been heterozygous (*Bb*).

The inheritance of human blood groups

This is still an example of monohybrid inheritance, but this time, there exist three possible alleles, only two of which are possessed by any one person.

The alleles are: I^A, I^B and I^o.

Both I^A and I^B are dominant over I^o, but are **codominant** to one another. Thus, the following combinations resulting in the blood groups, as shown, are possible:

genotype	phenotype
$I^A I^A$	blood group A
$I^A I^o$	blood group A
$I^B I^B$	blood group B
$I^B I^o$	blood group B
$I^o I^o$	blood group 0
$I^A I^B$	blood group AB

Example

An example of the inheritance of blood groups in humans can be shown in the following genetic diagram:

Parents:

genotypes	$I^A I^o$	$I^B I^o$
phenotypes	group A	group B

Gametes: I^A I^o I^B I^o

	I^A	I^o
I^B	$I^A I^B$	$I^B I^o$
I^o	$I^A I^o$	$I^o I^o$

Offspring:

Possible genotypes	$I^A I^B$	$I^A I^o$	$I^B I^o$	$I^o I^o$
Phenotypes (blood groups)	AB	A	B	O
probability	25%	25%	25%	25%

Sex-linked characteristics

Some characteristics are very much more common in men than in women. These are referred to as **sex-linked characteristics**. About 8% of males are colour blind, while far fewer than 1% of females have the condition.

The inheritance of colour-blindness

Sex-linked conditions are controlled by genes that are found on the **X chromosome**. Thus females, with two X chromosomes, will possess two alleles controlling their normal colour vision. The colour vision of men, having only one X chromosome, will depend on the inheritance of only one allele.

Let N be the dominant allele responsible for normal colour vision, then its recessive partner will be n.

Thus, a woman with normal colour vision will have the genotype NN or Nn and a man with normal colour vision will be N - (- indicates that the Y chromosome does not possess an allele for this characteristic).

If a woman is heterozygous (Nn), she will still have normal colour vision, but she is said to be a **carrier**. If a carrier female and a normal male have a child, the following diagram shows the possible genotypes and phenotypes of that child:

(NB The male gamete that carries the Y chromosome does not possess an allele for normal colour vision, therefore such a gamete is represented as '-' in the diagram.)

Parents:		mother × father	
		carrier	normal
Genotype:		Nn	N -
Gametes:	N n	N -

	N	n
N	NN	Nn
-	N -	n -

The child could be either: NN (a normal female) or Nn (a carrier female)

N - (a normal male) or n - (a colour-blind male).

These possibilities are equally likely – that is, 1 : 1 : 1 : 1.

The definition of a **sex-linked characteristic** is therefore **a characteristic in which the gene responsible is located on a sex chromosome making it more common in one sex than in the other**.

The Y chromosome carries very few genes, but is known to be responsible for some sex-linked conditions. One allele carried on the Y chromosome causes a particular form of baldness. Since only males have the Y chromosome, only males can have this particular form of the condition.

Chapter summary

■ You have learnt the difference between the term gene and allele.

■ You have learnt that the sex of a person is the result of the presence of X and Y chromosomes (not X and Y genes).

■ You have learnt the difference between mitosis and meiosis and are able to spell both words.

■ You have learnt the difference between genotype and phenotype and between homozygous and heterozygous.

■ You able to draw a full genetic diagram, including the words parents, gametes, F_1 or offspring and/or F_2, and showing the expected ratios.

Exam-style questions

1 Copy and complete the sentences by writing the most appropriate word in each space.

Use only words from the box.

alleles	diploid	dominant
gametes		
genotype	haploid	heterozygous
homozygous		
meiosis	mitosis	phenotype
recessive		

Height in some plants is controlled by a single pair of genes that has two _____, one for tallness (T) and one for shortness (t). Male _____ inside pollen grains are produced by a type of cell division known as _____, which ensures that nuclei are _____. When a _____ tall plant (Tt) is crossed with a short plant, half the offspring are tall and half of them are short. The _____ of the short plants must have been _____ [8]

2 a Colour blindness is a sex-linked condition. Explain the fact that most colour blind people are male. [4]

   b Copy and complete the genetic diagram to show how parents with normal colour vision can give birth to a child who is colour blind.

                              mother/father*   mother/father*
   parental
   genotypes

                                    ×

   genotypes
   of offspring
   phenotypes
   of offspring

   *delete as appropriate.

                                                    [5]
                                              [Total 9]

3 If a man who is heterozygous for blood group A has a child with a woman who has blood group AB, use a genetic diagram to calculate the chances of their child having blood group B. [8]

Variation and natural selection

19.01 Variation

Variation is the **difference between individuals of the same species**.

Sexual reproduction leads to variation in the offspring. No two offspring from the same parents, produced by sexual reproduction, are genetically identical (unless they have developed from the same ovum and sperm, in which case they are 'identical twins' and even then they will show very minor differences).

Two types of variation are seen, as shown in Figure 19.1:

1 **Continuous variation** that results in a **range** of phenotypes between two extremes – for example, height in humans

2 **Discontinuous variation** where there are a limited number of phenotypes with **no intermediates** – for example, the ability to roll your tongue into a 'U' shape – either you can or you cannot do it!

Continuous variation

Continuous variation in phenotye is the result of the interaction of two factors:

1 The genes that are inherited by an individual.

2 The effect of the environment on the individual.

The environmental factors involved might include:

1 The availability and type of food (in animals)

2 Disease

3 The climate – amount of sunlight, temperature and amount of available water

4 The ions present in the soil (in plants)

5 Competition from other organisms in the environment.

In continuous variation, individuals show a **range between the two extremes**. Every possible intermediate will exist.

Discontinuous variation

Discontinuous variation, except in some very extreme cases, is **caused by genes alone and results in a limited number of distinct phenotypes with no intermediates**.

Example

Blood groups – you are Group A, B, AB or O. (There are, however, several 'types' of these four groups – such as either Rhesus positive, or Rhesus negative.)

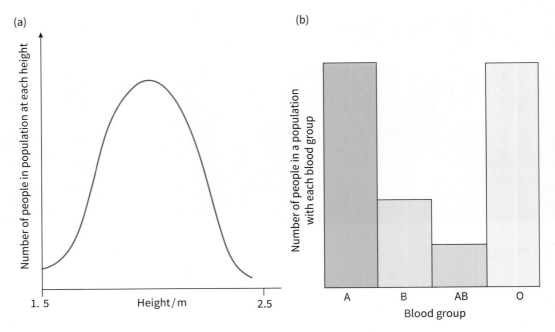

Figure 19.1 (a) Continuous variation (with every possible intermediate between two extremes); (b) Discontinuous variation (no intermediates)

Variation as a result of mutation

Genes and chromosomes are always subject to change (or mutation) as a result of environmental forces acting upon them. These forces are known as mutagens and include X-rays, atomic radiation, ultra-violet light (i.e. **ionising radiation**) and **some chemicals**. Exposure to higher doses of any of these mutagens will lead to a greater rate of mutation and thus a greater range of new, mutated, alleles.

A **mutation** is a **change in a gene (or chromosome)**.

This change in a gene – that is, a **gene mutation** – comes about as a result of a **change in the base sequence of DNA**.

An example of a gene mutation is that which causes sickle-cell anaemia. Hemoglobin, the oxygen-carrying pigment in red blood cells, is a protein. However, the gene for hemoglobin production can undergo a mutation in which one of the bases in one of its DNA triplets is changed. The hemoglobin produced thus has one incorrect amino acid in its chain and no longer carries hemoglobin efficiently.

Sickle cell anaemia is an example of monohybrid inheritance. The dominant allele is necessary before a person can make normal, functional hemoglobin in their red blood cells. If two parents who are heterozygous for the condition (and therefore have a 'normal' phenotype) both pass on a mutated (and recessive) allele to their offspring, the child is thus **homozygous recessive** and suffers sickle cell anaemia with the following problems:

- They are unable to make normal hemoglobin in red blood cells.
- Their red blood cells cannot carry sufficient oxygen.
- Their red blood cells have less flexible cell membranes and take on a distorted shape ('sickle-shaped').
- The deformed cells may block capillaries and encourage thrombosis.
- Limbs may swell and severe pain and aching may be experienced almost anywhere in the body.
- The patient may feel tired and experience shortness of breath and heart palpitations.
- The person is likely to die at an early age.

However, there is an **advantage** to being **heterozygous** in areas where **malaria** exists, as such individuals have a **resistance** to the disease and survive in high numbers. This survival leads to greater numbers, in regions where malaria exists, of offspring from heterozygous parents and thus to a greater chance of offspring being homozygous recessive and therefore born with sickle cell anaemia.

In genetic diagrams, the alleles are usually represented by Hb^A for the normal dominant allele and Hb^S for the recessive allele.

Parents who are both heterozygous for sickle cell anaemia have a 1 in 4 chance of producing a child with the disease.

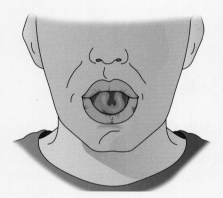
Adaptive features

The fact that organisms naturally show variation and that they are able to hand on that variation to their offspring has allowed them to adapt to a wide variety of different habitats.

Adaptive features are inherited features that help an organism to survive and reproduce in its environment.

Adaptations of leaves, stems and roots to different environments

Depending on where a plant grows, so may the organs of that plant can be adapted for life in the plant's particular environment.

Characteristics that are inherited will often be in the form of features that will help an organism to survive in the environment in which it lives. Such features are called adaptive features. An **adaptive feature** is defined as follows: **an inherited feature that helps an organism to survive and reproduce in its environment**.

Note that if an organism is able to survive in its environment, then there is a high chance that it will be able to reproduce.

Giraffes, for example, have long necks that enable them to reach food from trees long after shorter-necked leaf-eaters have eaten most of the leaves at lower levels. The giraffe remains well-fed, healthy and readily breeds more of its kind. It also has a mottled pattern of hair on its skin, allowing it to be well camouflaged while feeding amongst the trees – another adaptive feature.

Adaptations shown by plants that live in water (hydrophytes)

Light will penetrate only a limited distance in water, thus the leaves of flowering plants need to be near the water's surface. In the water lily (a **hydrophyte**; Figure 19.3) this is achieved as follows:

- Broad flat leaves use the water's surface tension to float of the water surface.

- Stomata on the upper surfaces of the leaves allow more efficient gaseous exchange.

- Plentiful air chambers within the mesophyll of the leaf help to give it buoyancy.

- Long flexible stalks allow the leaves to reach the water surface and allow movement without damage, while the **roots** can penetrate the mud at the bottom of the pond.

- The broad leaves reduce further the amount of light penetrating the water that could be used for photosynthesis of other plants competing for nutrients (more dilute in pond water than in soil).

Figure 19.3 Water lily

Adaptations shown by plants that live in dry places (xerophytes)

In dry places, such as deserts, rainfall is low, temperatures are high and thus, so too, is the tendency for evaporation.

Cacti (Figure 19.4) are **xerophytes** and are adapted as follows:

- Leaves are reduced to thorns – as a protection against herbivores, to reduce the number of stomata and thus to reduce also the amount of transpiration.

- Stems are green to carry out the photosynthesis that would otherwise be performed by the leaves.

- Stems are also often pleated so they can swell to store water during the brief period when rains occur.

- Roots are long so they can penetrate to the water table deep in the soil.

Figure 19.4 Cactus

The degree to which an organism is able to survive and reproduce in its environment is known as its **fitness**. An adaptive feature will function in a particular way to improve the organism's chances of survival. Those organisms with the greatest fitness will be most likely to survive and reproduce. The definition of adaptive feature given previously can therefore be extended to incorporate the idea of fitness so that an **adaptive feature is any inherited functional feature of an organism that increases its fitness. Fitness is the probability of that organism surviving and reproducing in the environment in which it is found**.

Progress check 19.2

Using the list of organisms you made and their stated adaptive features, for each feature you mention, explain how you think the feature increases the organism's fitness.

19.02 Selection

Natural selection

If, as a result of variation, the phenotype of an individual is such that it is at an advantage in a given habitat over others of that species, it is more likely to survive (they are said to be **better adapted** to their environment). Survivors then produce many offspring and hand on the genes that provided them with the advantage. Thus, some of their offspring will inherit the advantage – some may even improve on it. These offspring are thus also more likely to survive, until most of the population possess the 'advantage' that becomes an advantage no more, and the process then follows a similar course but this time perhaps focused on some other advantageous characteristic.

Examples of the variation shown by some members of a population in a given habitat might include:

- The shade of colour of a leaf-eating insect,

- The sharpness of bird of prey's vision and

- The speed at which a gazelle can run.

In all these examples, the variation can have some effect on the success or even on the chances of survival of that organism in its environment. All members of the same species will be competing with one another for basic resources such as food. This leads to what is often termed a 'struggle for survival'. Those best equipped for that struggle will be those that are better adapted and they are likely to survive, have the largest number of offspring and hand on the alleles for that particular advantage to at least some of their offspring (who, as a result of variation, may be even better adapted for survival – etc.). This type of selection of the best adapted (fittest) is called natural selection.

The better-camouflaged leaf insect may escape the notice of a hungry predator; the sharper the vision of a bird of prey, the more likely it is to find a meal – particularly important when food is scarce; the faster the gazelle can run, the more chance it has of escaping with its life from a hungry lion.

All organisms are therefore in competition with other members of their species in that particular environment. The winners in that competition survive to reproduce. The losers provide food for predators or fail to obtain enough food to remain alive.

It is the environment that 'decides' which organisms survive. The refinement of the adaptive features happens by chance mutations – never because the organisms make any attempt to adapt. The process can be explained by natural selection.

Since the variation is controlled by the alleles possessed by the organisms, the surviving individuals are able to hand on their advantages, via their alleles, to the next generation.

Worked example

a An owl is a bird of prey that hunts mice at night. Explain how natural selection has led to improved night vision in the owl and describe the advantages of improved night vision to the owl.

b Suggest ways in which improved night vision in owls may have had an effect on the local population of mice.

Answer

a In dim light, an owl requires sharp vision to see its prey. Any mutation affecting the development of night-vision cells in the retina will affect the ability of the owl to see at night. Owls breed by sexual reproduction. Sexual reproduction always produces variety, thus some owls will be born with slightly inferior night vision and some will have slightly better night vision. Those with the better night vision will be able to see the mice in dim light and thus be able to feed themselves and their young. They and the young will survive, while those with the poorer night vision may not catch enough food and die. Thus, only those with the better night vision will survive and they will hand on their better night vision to their offspring – some of whom, by variation caused by further mutation, may have night vision that is better still. This process occurs over many generations and thus owls' night vision is constantly becoming sharper. The advantage of this is that improved night vision will ensure that some owls will remain well fed, healthy and able to breed successfully.

b Mice also breed sexually and thus show variation as a result of mutation. Such variation may be in the colour of their fur, which makes them more difficult to see, especially in dim light. Some may develop improved hearing, making them that bit more able to hear an approaching owl. Such mice survive. Those that do not have these advantages are more likely to become food for the owl, those that possess the advantages survive, breed and hand on the advantages to their offspring – which again show variation. Thus evolution of both owls and mice run parallel and continuously, with each one helping to direct the evolution of the other.

(In a question of this type, it is important to describe the sequence of events in order and to mention that the process takes time, many generations, since each individual change is usually very small. It is important, also, not to suggest that the owls or the mice need to change or that there is any effort on the part of either to do so. It is something that happens purely by chance.)

As a result of natural selection, a population of organisms gradually becomes better adapted to survive in a given habitat. But habitats are always changing, so the process of natural selection is always ongoing and so, too, is the process of adaptation to that changing environment.

The process of **adaptation** is defined as **the process, resulting from natural selection, by which populations become more suited to their environment over many generations**.

Adaptation by natural selection has led to the process we know as evolution.

Adaptation of pathogenic bacteria to antibiotics

When antibiotics were first used to counteract pathogenic bacteria, they were very successful at destroying the targeted bacteria. However, bacteria show variation and, within any patient, some of the pathogenic bacteria possessed more resistance to the antibiotic than others and took longer to be destroyed. If the patient stopped taking the antibiotic course before all the bacteria were killed, the better-adapted resistant strain remained alive to be passed on to other hosts. This time:

- It would take longer for the same antibiotic to destroy the bacteria.

- The bacteria may become completely resistant to the antibiotic, which would now be of no use in controlling the infection.

- A new antibiotic would be needed, which would be expensive to produce and take some years to pass all the necessary safety tests.

It is vital for a patient to take the **entire course** of prescribed antibiotics and not stop just because they feel better. In this way, all the pathogenic bacteria will be destroyed leaving none of the fitter ones to be passed on. Prescription of antibiotics for minor bacterial infections should be discouraged.

19.03 Artificial selection

When it is not the environment, but humans who are doing the selecting, then the process is called **artificial selection** or **selective breeding**.

Many crop plants and farm animals of today are the result of selective breeding programmes. By selecting farm animals or crops that have desirable qualities and then using them for breeding, we can expect that most of the offspring will have the features of their parents, some may be even better, and these will be selected for further breeding.

Progress check 19.3

1   What is the value of adaptive features?

2   List as many organisms, plant and animal, as you can and state where they live and any adaptive features you think they have to help them survive and breed in their environments. Animals mentioned in Chapter 1 may help you.

The greatest use of artificial selection is in agriculture.

Examples include:

- Increased milk production in cows.

- Increased meat and better quality wool production in sheep.

- Increased yield from cereals.

- Increased disease resistance in many crops.

As a result, greater profits are made from greater quantities of better quality produce. In all cases, the following procedure is followed:

1   The individuals showing the quality required are selected.

2   Those individuals are used as breeding stock.

3   From their offspring, only those showing the desired quality to the greatest extent are selected.

4   These selected individuals are used for breeding.

5   This process is continued over many generations.

There is a danger, however, that this form of 'inbreeding', will increase the chances of some undesirable genetically-controlled deformity arising.

Artificial selection may thus be considered to be the production of varieties of animals and plants with increased economic importance.

The significant differences between natural selection and artificial selection are shown in Table 19.1:

Natural selection	Artificial selection
has been going on for many millions of years	has been going on only since humans have been on the planet
the result of natural environmental forces	tailored by humans to suit their needs
a relatively slow process	a much quicker process
offspring healthier as there is no inbreeding	inbreeding produces weaker individuals (prone to disease and genetic disorders)

Table 19.1

Chapter summary

■ You have learnt about the two types of variation – continuous and discontinuous.

■ You have learnt how variation fits organisms to their environments.

■ You have also learnt how variation can lead to selection – both natural and artificial.

Exam-style questions

1 Giving an example of each, distinguish between continuous and discontinuous variation. [6]

2 Explain the ways in which the allele that is responsible for sickle cell anaemia can be either and advantage or a disadvantage to a person who possesses it. [8]

3 Describe the similarities and differences between natural and artificial selection (selective breeding). [10]

Organisms and their environment

Learning outcomes

By the end of this chapter you should understand:

- ■ How energy flows through biological systems – food chains and food webs

- ■ Pyramids of numbers and of biomass

- ■ Natural cycles – of carbon, water and nitrogen

- ■ Populations and the control of their growth

20.01 Energy flow

The source of energy for all biological systems on the planet is **the Sun**. Living organisms use that energy for their activities and, so long as the Sun continues to shine, so long can life continue.

Not only is the (heat) energy from the Sun used by organisms such as 'cold-blooded' animals to keep their bodies warm, the Sun supplies the (light) energy which is locked away by plants during the manufacture of carbohydrates by photosynthesis. This energy is then passed on to animals that eat (ingest) the plants (or to animals, which eat animals that have eaten plants). In this way, the Sun's energy enters, then flows through biological systems. Note energy flows through biological systems. It is used up by the organisms involved or released into the environment. It is **never recycled**.

20.02 Food chains and food webs

Starting with the plant (the **producer**) that locks the light energy away as it photosynthesises, the energy then passes through successive organisms that make up a particular **food chain** as each organism ingests the one that comes before it in the chain.

Thus a **food chain shows the transfer of energy from one organism to the next beginning with a producer.**

Example

An example of a food chain is given here:

mahogany tree → caterpillar → song bird → hawk

TIP

Include at least three organisms in the chain and link all organisms mentioned with arrows – the arrow head pointing in the direction in which energy flows through the chain.

Progress check 20.1

1   With which type of organism do all food chains begin?

    A   carnivore

    B   decomposer

    C   herbivore

    D   producer

2   See if you can give two or three food chains for the area in which you live, and also some food chains for other regions such as a pond or lake, the sea and, perhaps, a desert or polar region.

We need to consider what happens to energy as it flows through a food chain.

- It enters the producer as light energy.

- It is locked away within the producer as chemical energy.

- It passes from organism to organism in the food chain as chemical energy. Each successive organism is said to be at a different **trophic level**.

- A **trophic level** is the **position of an organism in a food chain, food web, pyramid of numbers or pyramid of biomass**. (See sections on pyramids of numbers and of biomass.)

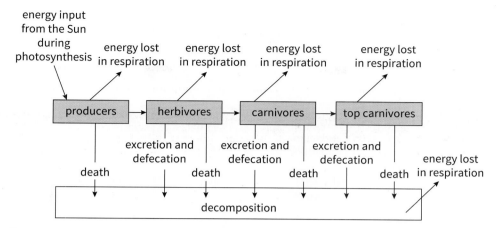

Figure 20.1 Energy flow in an ecosystem

- Each organism unlocks some of this energy to **use** for various processes within its body, for example, making new cells and the large organic molecules within them (during growth), muscular contraction (and movement), generating electrical impulses in the nervous system and raising body temperature. The chemical reaction that unlocks the chemical energy for conversion into other forms is respiration.

- Energy is used up in most of these processes. Only in the form of heat from an organism's body is it released to the environment outside the food chain. This includes energy from the respiration of bacteria and fungi that eventually decay dead organisms.

> **TIP**
>
> Avoid saying that energy is 'needed' for respiration (it is **released** by respiration).

Much of the energy is still present in the faeces and some in the nitrogenous waste of animals. This energy is available to **decomposers**. Not all **herbivores** are eaten, thus the amount of energy left within herbivores to be passed on to **carnivores** is small – 20% (only 2% of the original amount in the producer).

For this reason, food chains are limited in length, as there is insufficient energy remaining to sustain a succession of carnivores. **Five** trophic levels are usually the limit for a food chain (Figure 20.1).

The longer the food chain, the less the energy available to the top carnivore at the end of the chain. Short food

chains are therefore much more energy efficient than long ones. In order to supply enough energy in food to maintain an ever increasing world population, it must be realised that far less energy is lost when man eats green plants than when crop plants are fed to animals, which are then eaten by a human.

In any one habitat, such as a pond or mangrove swamp, there will be many organisms living together. In some way they will all be interconnected by way of different food chains.

A **network of interconnected food chains** is known as a food web (Figure 20.2).

While all food chains (and thus all food webs) begin with a producer, food webs may begin with several different species of producer.

A **producer** is an organism that makes its own organic nutrients, usually using energy from sunlight, through photosynthesis.

In a food chain or food web, producers are eaten by consumers.

A consumer in a food chain is an organism that gets its energy by feeding on other organisms.

An animal that gets its energy by eating plants is an herbivore (or primary consumer).

An animal that gets its energy by eating other animals is a carnivore (or secondary consumer – the consumer that feeds on the secondary consumer is a tertiary consumer – and so on). Thus all consumers above the level of herbivore, that is, all meat eaters, are carnivores.

When all organisms in a food chain or web die they are decomposed largely by bacteria and fungi.

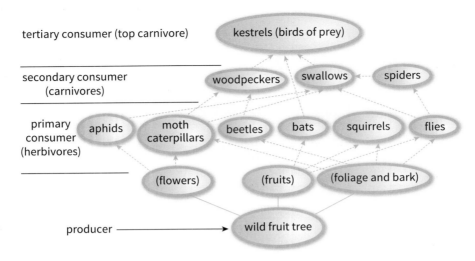

Figure 20.2 A food web

Decomposers are organisms the get their energy from dead or waste organic material. They release enzymes to break down large molecules in dead organic matter into smaller ones that can then be recycled.

The food web in Figure 20.2 shows that if fruits from wild fruit trees are harvested by humans, bats would be affected as they would suffer a shortage of food. Woodpeckers, swallows and spiders would then be eaten in larger numbers by the kestrels in order to make up for the lack of bats.

Progress check 20.2

1   If the fruit from the tree was harvested by humans, what do you think might happen to the aphids, moth caterpillars, beetles and flies?

2   What do you think might happen if a fourth type of secondary consumer was introduced into the area?

Pyramids of numbers

- In order to provide enough food (and therefore enough energy) for their own metabolic processes, to provide enough food for the herbivores that eat them and leave enough surviving individuals to produce the next generation, we would expect there to be a larger number of producers (plants) than primary (1st) consumers (herbivores). For the same reasons, we would expect there to be more primary consumers than secondary (2nd) consumers, more secondary than tertiary (3rd) and more tertiary than quaternary (4th) consumers – if the food chain

is sufficiently long enough to include quaternary consumers. These decreasing numbers along a food web can be represented in the form of a **pyramid of numbers** as shown in Figure 20.3.

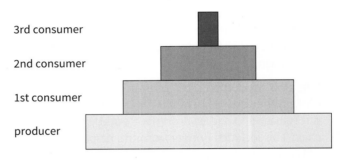

Figure 20.3 Pyramid of numbers for a food web

This representation of feeding relationships can be misleading, however, since one large plant may sustain a large number of very small herbivores (as shown in Figure 20.2 of the food web). One herbivore may sustain a large number of small carnivores. This then leads to a 'top heavy' pyramid as shown in Figure 20.4.

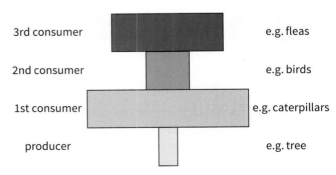

Figure 20.4 'Top heavy' pyramid of numbers for a particular food chain

Overharvesting and the introduction of foreign species

In an environment that receives relatively little human interference, the organisms in a food chain or food web reach stable numbers. Organisms at each trophic level have sufficient food for their needs without causing a permanent steady decline in the numbers of the organism at the previous trophic level.

However, if there is human interference and one trophic level is overharvested, for example, selected species of food fish in the sea or trees in a forest, then those organisms that rely on these species as a food source will suffer and may even become extinct.

The introduction of species not native to an environment can also have disastrous consequences. In 1935, the cane toad was introduced into Australia (from South America) to destroy the sugar cane beetle that was affecting sugar crops. However, the cane toad has depleted the populations of other less harmful insects. In doing so, they have removed important links in locally established food chains and food webs, causing a decline in the numbers of some lizards, snakes and native toads. Toads have poison glands in their skin and local mammals, including pets, have been poisoned. Furthermore, there is no good evidence that the sugar cane beetle has be adversely affected at all!

Pyramids of biomass

A **pyramid of biomass** is a pyramid constructed using the dry mass of organisms at each trophic level in a food chain (or food web). It produces pyramids of a more familiar shape (as in Figure 20.3) and can be constructed by collecting data from population estimates in any particular habitat. Biomass is the total dry mass of a population – that is, the theoretical mass of chemicals other than water in the organisms under consideration (water being liable to considerable variation). Thus it would take account of the total mass of leaves produced by the tree and give a more realistic picture of the amount of food available to the herbivores (and also of the amount of energy contained in it).

Progress check 20.3

1   For one of the food chains you drew earlier, add further organisms so that your food chain forms part of a food web.

2   The diagram shows a food chain.

    Tree → insect larva → small bird → bird of prey

    Which will be represented by the narrowest block in a pyramid of numbers?

    A   bird of prey

    B   insect larva

    C   small bird

    D   tree

3   The diagram show a food chain.

    Tree → insect larva → small bird → bird of prey

    Which of the organisms in the food chain will be represented by the widest block in a pyramid of biomass?

    A   bird of prey

    B   insect larva

    C   small bird

    D   tree

20.03 Nutrient cycles

The carbon cycle

The carbon cycle is shown in Figure 20.5. When **decomposition** occurs, carbohydrates in the dead organic matter are used as the substrate for respiration by decomposers (bacteria and fungi). Carbon dioxide is released into the atmosphere, as it is in all cases of aerobic respiration. **Combustion**, the burning of fuels, many of which are fossil fuels such as coal, gas and oil, also releases carbon dioxide into the atmosphere.

Photosynthesis absorbs that carbon dioxide and converts it into carbohydrates (which may then be used to make fats or proteins) in a plant. Animal nutrition (**feeding**) involves the transfer of these carbon-containing molecules to the animals, which, along with the plants, respire or die and are decomposed, releasing the carbon dioxide back into the atmosphere and so the cycle continues.

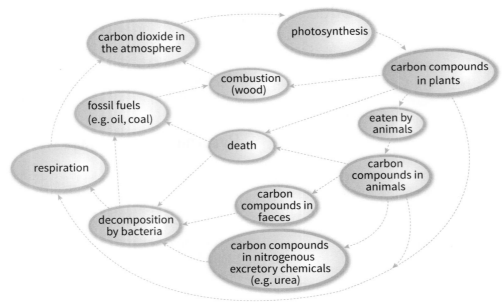

Figure 20.5 The carbon cycle

Human interference with natural cycles

The burning of fossil fuels (coal and gas) releases carbon dioxide into the air. The cutting down of rainforests (on a massive scale in some areas such as the Amazon basin), decreases the amount of carbon dioxide being absorbed for photosynthesis. Levels of carbon dioxide in the atmosphere are increasing. It is widely believed that the extra carbon dioxide in the atmosphere forms a 'blanket' around the earth (the **greenhouse effect**) reducing the escape of heat and leading to the phenomenon called **global warming**. This could lead to the melting of the ice caps, and possible subsequent flooding.

Deforestation also reduces the amount of water lost to the atmosphere through transpiration, which may have an effect on the amount of rainfall in regions that may be at some considerable distance from the destruction. Conversely, burning fossil fuels increases the amount of atmospheric water.

The importance of water and the water cycle

The water cycle is shown in Figure 20.6.

Water is a chemical that is essential to life on this planet. It is important for the following reasons:

- It is the medium in which **all chemical reactions** take place.

- Water is often called the **universal solvent**.

- It is the means of **transporting** chemicals in plants and animals.

- It is used in **temperature regulation** in many animals.

- It is a **major constituent** (about 85%) of all cells.

Water is used during photosynthesis and released during respiration. These two processes, thus, play a part in the recycling of water as well as in the carbon cycle. However, there are processes that are involved only in the water cycle. These are:

- **Transpiration** – the loss of water vapour from plants

- **Evaporation** – from soil, seas, rivers, lakes, ponds and sweat

- **Condensation** – the concentration of water vapour to form clouds

- **Precipitation** – rain/sleet/snow.

TIP

You are advised to treat transpiration (from plants) and evaporation from soil and bodies of water as two separate processes, though they are sometimes considered as one process under the term evapotranspiration.

After evaporation, water vapour rises and condenses to form clouds that then lose their water.

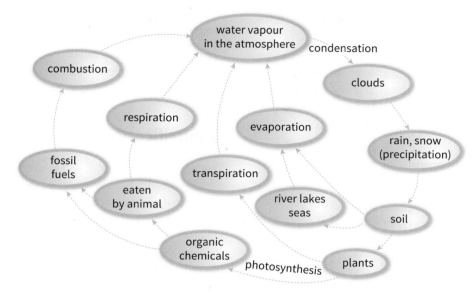

Figure 20.6 The water cycle

Worked example

Explain the effects on the water and on the carbon cycle of deforestation and combustion of fossil fuels.

Answer

It would be wise to take each part of the question separately. First, the effect on the water cycle of deforestation. A forest is a large expanse of woodland. To remove these trees will reduce transpiration. Less water vapour will be released into the atmosphere, reducing cloud formation and therefore reducing rainfall. However, the amount of rainfall is unlikely to be greatly affected in the region of deforestation, since wind carries clouds many miles before they release their rain. The effect on the carbon cycle of deforestation will be a decreased rate of carbon dioxide uptake for photosynthesis. It will thus create a build-up of carbon dioxide, which is a greenhouse gas that leads to global warming. There would also be a decrease in the amount of carbon dioxide released by the trees during respiration, but over a 24-hour period, plants photosynthesise more than they respire (otherwise they could not make enough food to grow).

The effect on the water cycle of burning fossil fuels is to release water vapour into the atmosphere. This would increase cloud formation – and then rain, but only to a very small extent. The effect of burning fossil fuels on the carbon cycle is more significant, since carbon dioxide is a major product of combustion. The industrial burning of fossil fuels is a major contributor to increased global warming.

Note that there may be a temptation to talk about the pollution caused by burning fossil fuels, but it is important to answer the question set, which requires mention only of the water and carbon cycles.

The nitrogen cycle

The nitrogen cycle is shown in in Figure 20.7.

Decomposition includes the conversion by bacteria of proteins to amino acids and the conversion of amino acids and urea to ammonium ions. This process usually takes place in the soil.

Urea is present in urine from animals where it is made in their livers by a process called deamination. During deamination, (excess) amino acids are broken down into two chemicals – a carbohydrate that is stored and the nitrogen-containing part (urea) that is excreted.

Ammonium ions contain nitrogen atoms, but before plants can absorb nitrogen from the soil, it must be in the form of soluble nitrate ions.

Nitrifying bacteria convert ammonium ions first into nitrites then into nitrates (during the process called nitrification).

Plants **absorb the nitrates** and use them together with the carbohydrates made by photosynthesis, to make **amino acids** and then **proteins**.

Animals eat the protein in the plants and excrete, for example, urea. Also, both plants and animals die. Decomposition releases ammonium ions again.

Atmospheric nitrogen (78% of the air) cannot be used either by plants or by animals in its gaseous form, but some bacteria (the **nitrogen-fixing bacteria**) can do so. Such bacteria are found in two forms:

1 Those that live in swellings called **nodules** on the roots of peas, beans (and other leguminous plants).

2 Those that live freely in the soil.

Nitrogen fixation by these bacteria changes the atmospheric nitrogen, via ammonia, into proteins and becomes available to other organisms when these proteins later decompose.

Nitrogen fixation also occurs when **lightning** passes through the nitrogen in the air, converting it to nitric acid, which forms nitrates in the soil.

Denitrification

The action of nitrogen-fixing bacteria would, eventually, deplete the nitrogen content of the air. Thus, there are **dentrifying bacteria** that use nitrates in their metabolism, releasing gaseous nitrogen as a waste product. They live in infertile soils lacking in oxygen (they employ anaerobic respiration) and further reduce soil fertility by removing nitrates.

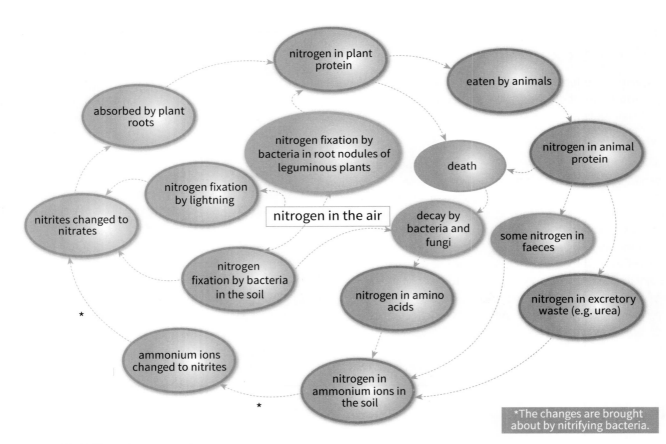

Figure 20.7 The nitrogen cycle

20.04 Population size

Within any habitat there will be many different species living together and relying on one another for food. All members of the same species form a population of that species in that habitat.

The definition of a **population** is a **group of organisms of the same species living in the same area at the same time**.

If a very few members of a new species arrive in a new habitat, as long as the habitat is suitable for them, they will survive, reproduce and establish a new population within the habitat. However, the size of that population will depend on the following factors:

- **Food supply** – There must be sufficient food in the new environment for all members of the population – and considerably more than is required by them if the population is to increase in size.

- **Predation** (the presence of predators that might eat them) – Some of the already established populations in that environment may find that the new species is a welcome prey (additional food source) and if the new species has no way of protecting itself against these predators, the species will not survive. It follows that a lack of predators will help the new population to increase in numbers.

- **Disease** – There may be pathogens in the new environment against which the already established populations have developed immunity. The new species may not have any natural immunity to such diseases, which may limit its growth or even eliminate it completely.

A habitat is where an organism finds its food and shelter and where it reproduces (i.e. where it lives). All the organisms in a habitat, plus the non-living part of the environment (light, soil, air, water, etc.), make up an ecosystem.

Thus, an **ecosystem** is **a unit containing the community of organisms and their environment interacting together**. If we consider just the living part of an ecosystem, then we are considering the **community** within that ecosystem.

A **community** is thus **all of the populations of different species in an ecosystem**. Examples are a lake that contains populations of different species of plants and animals, forming a community and all living in water, which is one of the non-living features of their ecosystem.

Progress check 20.4

1 Can you list the various components of the ecosystem formed by a decaying log?

The rate at which the population grows in a new environment depends on the continued availability of resources.

To start with, there will be a period of time (the **lag phase**) in the growth of the population during which the species is acclimatising to its new environment. This is followed by a rapid increase in population size (the **exponential** or **log** phase). Eventually the population numbers become stabilised at its optimum level for this particular environment (the **stationary** phase). The length of time that the population stays at this level will depend on the stability of the environment in which they are living, but a change in any of the factors within that environment may cause the size of the population to fluctuate. Any of the relevant factors could therefore become **limiting factors**. Lack of food, increase in predation or disease could wipe out the population completely – that is, its size falls to zero (the **death phase**). This growth curve is shown in Figure 20.8 and an explanation of each phase follows the graph.

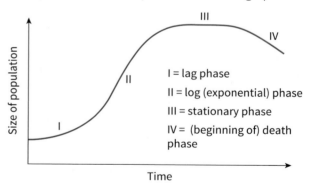

Figure 20.8 Curve of population growth

Because the first part of the graph is 'S' – shaped, it is said to be **sigmoid** ('sigma' is the Greek 'S').

The factors involved in the phases of population growth

Lag phase

New arrivals in a habitat must first find a suitable food source and, in a large area, those that rely on sexual reproduction may not readily find mates with which to breed. Few offspring are added to the population at first. The rate at which the population grows is therefore comparatively slow.

The exponential (or log) phase

Initially, one pair of organisms may produce, say, six offspring. The population would thus increase by six during one breeding season. However, sometime later, when say, there are 20 breeding pairs, then during one breeding season, the population would be likely to increase by 20 × 6 (= 120) in the same period of time. The same would apply to a single-celled organism reproducing by cell division. The first division will produce two organisms, but the tenth division will produce over a thousand organisms. The population is thus increasing at an ever faster rate. This is shown by the steep rise of the population growth curve.

The stationary phase

Eventually the environment is unable to sustain a larger population of this species. There is just enough food; predators and prey balance one another and only the weakest fall victim to disease. **Food, predation** and **disease** are all referred to as **limiting factors**, because any one of them may limit (and control) the size of the population. Climate may also play a part with temperature, rainfall and light intensity all being possible limiting factors.

The death phase

Particularly with organisms such as bacteria that decompose organic matter, a population may reach a maximum size, then relatively quickly fall to zero. This may be because the **food source** is **exhausted**. It may be due to the **build-up of excretory material** (e.g. alcohol when yeast is fermenting sugar) or competition with an organism newly introduced to the environment. Climatic (**abiotic**) factors may also be responsible.

Progress check 20.5

1   List five factors that may limit the size of a population.

2   When the size of an established population is being maintained by the naturally occurring factors in its environment, it is said to be in which phase?

    A   death

    B   lag

    C   log

    D   stationary

20.05 Population size

In a natural habitat, the rate of growth of a population and its eventual size within its balanced community depends on **three** basic factors:

1   The **amount of available food** – The more food available, the faster a population will grow and the larger it may become.

2   **Predation** – The number and efficiency of local predators will affect the population in that the greater the number of predators, the slower the population will increase in size and the smaller will be its final numbers.

3   **Disease** – The prevalence and severity of diseases within the population will have a controlling effect of the size and rate of growth of a population.

Human populations

While predation is unlikely to have an effect on the size of human populations, the other two factors will certainly have an effect. Over the last 250 years, there has been a very marked increase in the size of the human population. Figure 20.9 shows just how great it has been. If the population continues to rise at the rate shown in Figure 20.9, then there could be some serious problems arising as a result.

- Many houses, schools and hospitals will need to be built.

- That building may take place on land currently used for food production – reducing the availability of food required by the ever-increasing population.

- Moving food from where it can be produced to where it is needed is costly and the transport needed will add to the current problems of pollution.

- Some regions with a high population may not have sufficient water available.

- The planet may run out of its natural resources, especially fuel and metal ores.

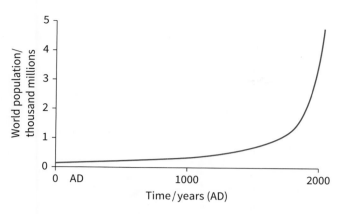

Figure 20.9 The graph of world population growth

The graph of world human population is effectively the lag phase and the beginning of the exponential phase of the growth of a population.

The human animal is as much part of the ecosystem in which he or she lives as any other organism in that ecosystem. But humans can be far more destructive than any other of their fellow organisms.

The global human population growth follows a similar pattern to that described for other organisms. However, it is still in the exponential phase. Human have been able to decrease the effect of a number of limiting factors to population growth with improved standards of nutrition, medical advances (antibiotics and surgery) and an absence of any predators have helped largely to push the limiting factors out of consideration. The birth rate greatly exceeds the death rate. Birth control may become very important if the human race is continue to be so little affected by the natural limiting factors to population growth.

Chapter summary

- ☐ You have learnt about food chains and food webs.

- ☐ You have learnt about pyramids of number and biomass.

- ☐ You have learnt about the carbon, water and nitrogen cycles.

- ☐ You have learnt about populations and population growth – and how it these are shown graphically.

Exam-style questions

1 a   Figure 20.10 shows a water cycle.

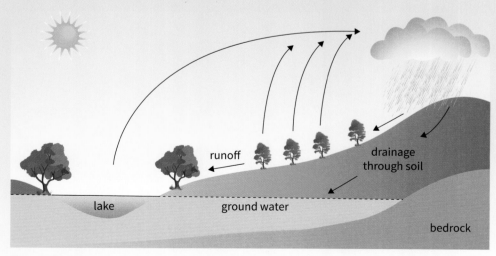

Figure 20.10

On a copy of Figure 20.10, mark clearly with the letters indicated where each of the following named processes is occurring:

letter	process
A	evaporation
B	transpiration
C	condensation

[4]

b   Suggest how the lake may become rich in nitrates. [2]

c   Explain why water is so important to living organisms. [4]

[Total 10]

2   Explain the phases in the growth of a population as it establishes itself in a new habitat. [12]

3   Figure 20.11 shows swellings on the root system of a plant. Explain how a farmer can improve the fertility of his soil by growing this crop.

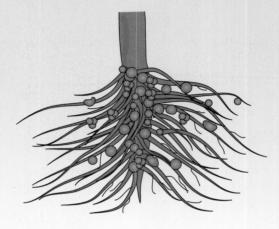

Figure 20.11

[8]

Chapter 21

Biotechnology and genetic engineering

Learning outcomes

By the end of this chapter you should understand:

■ Everyday and commercial use of biotechnology

■ How an understanding of how genetic structure has allowed us to select genes, and sometimes modify them, then use them to our advantage – often in disease resistance

21.01 Biotechnology

Biotechnology is the use of living organisms to make useful products and genetic engineering is the alteration of the structure of genetic material in a living organism.

Bacteria are often used in these processes because they are able to reproduce quickly (once every half-hour in suitable conditions) and their DNA can be used to manufacture complex molecules (see later).

Bacteria are ideal organisms to be used in biotechnology and genetic engineering. Laboratory mammals would appear to have more relevance to genetic experiments that have human significance, but the structure of chromosomes and the genetic code they use is exactly the same in humans and in bacteria. It is the same in all other organisms as well, so results with bacteria are relevant to all organisms.

Bacteria reproduce quickly and in large numbers. Growing them is easy and their use avoids criticism from people who, for ethical and other reasons, believe that genetic experiments on animals and on plants – especially food plants – are not justifiable. Finally, bacteria possess rings of DNA, called **plasmids**, which are very suitable for genetic engineering techniques.

The use of yeast in biofuel production and bread making

Yeast is a fungus that is able to respire anaerobically (i.e. not in the presence of oxygen). During this process, yeast has been shown to be able to break down **carbohydrates** in plants (including cellulose) in order to obtain energy and one of the waste products is the alcohol **ethanol**. The other waste product is carbon dioxide. This ability is made use of in the production of **biofuels** where plant waste (a suitable 'substrate')

together with yeast and water are placed in larger vessels (**fermenters**). The plant waste provides the carbohydrate and the ethanol produced can then be purified and used as fuel.

In **bread making**, yeast is mixed with sugar and flour. This time the carbon dioxide produced is used to bubble through the dough, lightening its texture. This process takes several minutes to an hour or so (depending on the air temperature) and is called 'proving' the dough. The dough is then baked in an oven, which has the effect of increasing the volume of the air pockets in the dough, but also of evaporating and thus removing the alcohol and killing the yeast.

Commercial use of enzymes in the production of fruit juices

The cells of a plant are held together by a chemical substance called **pectin**. Pectin is also found within the layers of the plants' cell walls. Bacteria and fungi that decompose fruit release the enzyme **pectinase** that breaks down the pectin thus exposing a larger surface area of cell walls, and also destroying the pectin within the cell wall, speeding up the process of decomposition. A small percentage of the pectinase used in the food industry is obtained by culturing a colony of the correct species of bacterium, allowing the colony to decompose plant material and harvesting the pectinase it produces.

Fruit juice manufacturers thus add **pectinase** to fruit to separate the cells in the fruit that is being used. As the cells are then pressed, they more readily release the juice within. In some beverages, pectin can cause a cloudiness that is removed with the use of pectinase (available commercially from pharmacies and health-food shops, sometimes called 'pectolase').

The effect of pectinase on apple cells can be demonstrated as follows:

PRACTICAL

Apparatus: 2 beakers each large enough to hold a chopped-up apple

A knife

Clingfilm (enough to cover both beakers)

A water bath set at 40 °C

2 measuring cylinders (or 1 with means of washing and drying it)

Materials: 2 apples

A packet of pectinase

At least 10 cm³ water

Method:

Cut two apples into small cubes and place them in two separate beakers. To beaker 1, add 4 cm³ of pectinase (made up in accordance with the instructions on the packet of pectinase obtainable from biological suppliers) and to beaker 2 add 4 cm³ of distilled water. Cover the top of each beaker with cling film. Place both beakers in a water bath at 40 °C and leave for 25–30 minutes.

Then pour the liquid from each beaker through a cloth filter into separate measuring cylinders. Record the volume of each.

Result:

The liquid from beaker 1 will have a volume greater than 4 cm³. The liquid from beaker 2 will have remained at 4 cm³. Can you explain this result?

Conclusion:

The pectinase has separated the cells and damaged the cell walls of the apple allowing juice to pass from the cells into the pectinase solution increasing its volume. Again, can you explain this result?

Commercial use of enzymes in washing powders

Stains are often **protein**, **fat** or **carbohydrate** in origin. Traditionally, stubborn stains have been removed from clothing by boiling with a soap-based detergent. The boiling denatures proteins such as those in blood stains and the detergent helps to break up and dissolve the protein, but it is not always effective on stubborn stains. It is now common to add enzymes to detergent washing powders.

The enzymes are:

- **Proteases** – for digestion protein based stains such as blood.

- **Lipases** – for fatty stains caused by, for example, milk and eggs.

- **Amylases** – for removing carbohydrate-based stains such those caused by sugars that are not removed before they have dried.

- **Cellulases** – for removing damaged cellulose fibres of cotton fabric, giving a worn garment a smoother finish.

The effectiveness of biological washing powders compared with non-biological washing powders can be demonstrated as follows:

PRACTICAL

1   The effect of temperature on the enzymes in biological washing powder

Apparatus: 50 cm³ biological washing powder solution

50 cm³ non-biological washing powder solution

6 large test-tubes and means of labelling them

2 pairs of forceps

Stopclock, or view of a clock with a sweep seconds hand

White tile

3 water baths set at 0 °C (ice and water), 40 °C and 75 °C

6 pieces of blood-stained white cloth (approx. 1 cm × 1 cm)

1 piece of cloth similar to the other cloth, but left unstained

The washing powder solutions should be made up by adding 8 g powder to 500 cm³ lukewarm tap water. If the solution is already provided in liquid form in a bottle, then dilute the solution using 0.75 ml to 500 cm³ lukewarm tap water.

To prepare blood-stained cloth, the cut surface of a piece of liver should be rubbed over the surface of the pieces of white cloth **the day before the practical** and allowed to dry in the fridge.

Method:

To a depth of about 3 cm, add biological washing powder solution to 3 of the test-tubes (labelled 1–3), and non-biological washing powder solution to the other three test-tubes, labelled 4–6.

Place test tubes 1 and 4 in the water bath at 0 °C.

Place test-tubes 2 and 5 in the water bath at 40 °C.

Place test-tubes 3 and 6 in the water bath at 75 °C.

Leave all the test-tubes for 15 minutes to allow the solutions to reach the temperatures of the water baths.

Submerge one piece of blood-stained cloth into each test-tube and record the time.

Submerge the non-stained piece of cloth in a depth of 3 cm water in test-tube 7.

At 2–3-minute intervals, shake each test tube and observe the colour of each piece of cloth.

Write the numbers 1–7 on the white tile.

When one of the pieces of cloth in test-tubes 1–6 has become noticeably paler, record the time and transfer each piece of cloth to the white tile, next to the number corresponding to the test-tube it came from.

Expected results:

Pieces of cloth 2 and 6 are likely to be the whitest (use cloth 7 as a comparison).

2   Investigation of the effect of pH on enzyme action

The requirements are as in the experiment investigating the effect of temperature. However, this time, you need only one water bath set at 40 °C and only three test-tubes are needed. To one of the experimental test-tubes containing the biological washing powder, add one drop of very dilute hydrochloric acid, add nothing to the solution in the second and add one drop of very dilute sodium hydroxide to the third. Incubate all three test-tubes and note the depth of colour in the stain of the cloth in the three test-tubes over the same period of time. The cloth exposed to the enzyme alone should be the first to lose its stain. Depending on the concentration of the hydrochloric acid and sodium hydroxide used, the other two may show no change in the stain at all.

Progress check 21.1

1   Using information from Chapter 5, explain the results for both of these investigations.

2   How do you think these results might differ if the test-tubes containing the pieces of cloth were left in the water baths for double your recorded time?

3   If the name of a chemical ends in 'ase', what type of chemical is it likely to be?

The advantages of using enzymes in washing powders

1   Enzymes can be used that target specific stains making them much more efficient stain removers.

2   Enzymes usually have an optimum temperature well below that of boiling water. This:

   • avoids the hazards of scalding

   • saves energy

   • decreases fuel costs.

3   Enzymes manufactured artificially are usually non-toxic.

Lactose intolerance

An inherited condition called **lactose intolerance** makes some people unable to digest lactose (or 'milk sugar'). This can cause sufferers to experience vomiting, bloating and diarrhoea if they drink milk or eat dairy products. The solution is to drink lactose-free milk, or products made from it. Milk is made lactose-free by gently heating it with a solution of the enzyme **lactase**, which can be obtained from bacterial cultures. The enzyme converts the lactose into simple sugars that do not affect the nutritional value of the milk but avoid the unpleasant effects of lactose intolerance.

Commercial use of the enzymes in the manufacture of the antibiotic penicillin

Biotechnology is used in many industrial processes other than enzyme production, for example, in the manufacture of **antibiotics**. **Microorganisms** (fungi and bacteria) play a central role in their production. Since large quantities of end-product are required, the microorganisms involved are grown in very large containers called **fermenters** (Figure 21.1). A **maximum rate of growth** of the microorganism within the fermenter can be achieved by careful control of:

1   The **temperature**

2   The type and concentration of **substrate** (i.e. the substance on which the microorganism works)

3   The **oxygen availability**.

Thus fermenters are unaffected by climate, soil type or, in temperate regions, time of year, though they must be kept **sterile** to prevent the growth of unwanted species of microorganism that might contaminate the end-product.

Pathogenic (disease-causing) bacteria are known to be killed by certain chemicals (**antibiotics**) released by other microorganisms, especially fungi.

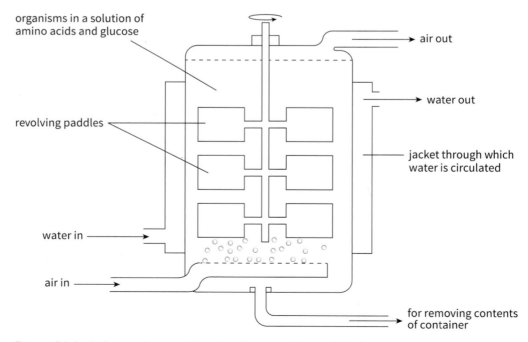

Figure 21.1 A fermenter used in manufacture of an antibiotic

The **mould fungus** *Penicillium*, which grows naturally on stale bread, is first cultured on a sterile medium such as corn steep in a laboratory.

The fungus is then introduced into a fermenter for commercial production. The fermenter (up to 100 000 litres in volume) contains a suitable sterile medium on which the fungus can grow.

After an appropriate period of time (when the rate of fungal growth becomes limited by the amount of food available), the contents of the fermenter are crushed and filtered. The liquid part contains the antibiotic (in this case, **penicillin**), which is separated and purified, while the rest of the material may be dried and used as cattle feed.

21.02 Genetic engineering

Recent advances in the understanding of genes and how they work have enabled scientists to manipulate them to carry out specific useful functions. The process involving this manipulation is called genetic engineering, which is defined as **changing the genetic material of an organism by removing, changing or inserting individual genes**.

As a result of genetic engineering, it is now possible to insert the human gene that makes the hormone **insulin** into bacteria. The result is that the genes of the bacteria not only make all the proteins that the bacteria require for their own metabolism, they also make insulin that can be extracted and used to treat diabetes.

By taking genes from plants that are resistant to herbicides and to insect pests, and inserting them into crop plants, the crop plants can also be made resistant to herbicides and insect pests — ensuring larger yields of better quality.

In countries where people tend to suffer from dietary deficiencies, crops are now being grown that have implanted genes that cause the plants to manufacture more vitamins.

Thus, apart from genetic engineering helping to treat disease, as in the example of insulin production, it is also helping to avoid diseases from lack of food or from dietary imbalance.

Worked example

Figure 21.2 shows the number of fungal 'cells' and the amount of penicillin they produce during commercial production in a fermenter.

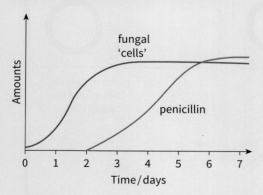

Figure 21.2

a   Name the fungus that is used and explain why the word 'cells' is not totally accurate for this organism.

b   i)   During which phase of the growth of the fungal population is the penicillin produced?

ii) Suggest a reason for this.

c   i)   Explain why the number of fungal 'cells' and the amount of penicillin produced become constant.

ii) Explain why it may not be possible to further increase the number of fungal cells or the amount of penicillin produced.

Answer

a   The question helps here as penicillin is made by the mould fungus *Penicillium*. Note that the genus is given. It starts with an upper-case letter and is in italics – or underlined if you are writing it.

Fungal hyphae are not divided into separate cells – they have nuclei scattered throughout their cytoplasm.

b   i)   You need to look back at your population growth graph (Figure 20.8 in Chapter 20) to see, on day 2 when penicillin starts to be produced, the fungal population has reached its stationary phase (always try to produce a relevant reading from the graph, and don't forget the units). ➔

ii) There is not enough fungus present at the start to manufacture recognisable amounts of penicillin. During the next 2 days, the fungus is using its energy for growth rather than for penicillin production.

c   i) Population growth is under the control of limiting factors. Such factors may be the temperature, the availability of food, or of

oxygen. Any one of these in short supply will limit the growth of the fungus and thus the amount of penicillin produced.

ii) If the limiting factor that is controlling these processes is at the optimum (best) for the fungus, then supplying more of it will not increase the amount of penicillin produced.

21.03 Genes and proteins

The nucleus of a cell controls the activities occurring in that cell's cytoplasm. In particular, it is the DNA in the chromosomes of the nucleus that controls the production of proteins in the cytoplasm. More precisely still, **each gene is responsible for producing one particular protein**.

Technological advances have enabled us to identify individual genes and to know precisely which proteins they are responsible for producing.

Gene transfer

A particular gene, even when transferred to another cell, will continue to operate as it would do in its original cell. Thus, it is possible to cause a cell to produce proteins of a type it would not normally produce, by introducing genes from the cells of completely different species.

Genetic engineering is thus defined **as taking a gene from one species and putting it into another species**.

This process has been shown to be of particular importance in the manufacture of the human protein insulin.

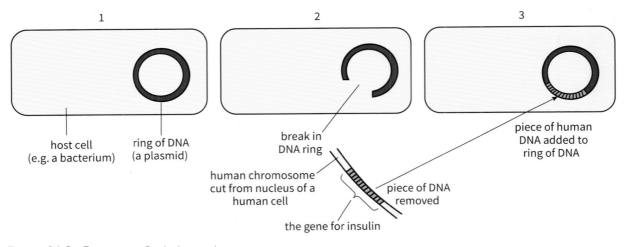

Figure 21.3 Gene transfer in bacteria

21.04 Commercial manufacture of insulin using genetic engineering

The gene for the production of human insulin (a protein) may be taken from a healthy person and inserted into a bacterial cell. The human DNA operates within the bacterial cell just as if it were in a human cell and causes the cytoplasm of the bacterium to manufacture insulin.

The process of gene transfer is shown in Figure 21.2; it is carried out as follows:

1   The gene for insulin production is 'cut' from the appropriate human chromosome that carries it. The DNA is broken precisely at the required points using enzymes known as **restriction enzymes**.

2   A circle of DNA found in a bacterium (commonly a bacterium called *Escherichia coli* or *E. coli*), known as a **plasmid**, is 'cut' open – again using the same restriction enzymes.

3   The human insulin gene is inserted into the cut plasmid. The joining ends of the human and the bacterial DNA, since they were cut with the same restriction enzymes, match in chemical structure (and are called **sticky ends**).

4   The sticky ends are joined this time using enzymes called ligases.

The 'genetically engineered' bacterium is then given optimum conditions for growth and reproduction (using a vat, similar to a fermenter, at a suitable temperature and containing a suitable substrate solution on which the bacteria can feed). The genes in the bacterial DNA not only manufacture all the proteins that the bacterium requires for reproduction and growth, but also the hormone **insulin**.

After a period of time, the solution (which contains the insulin) and the bacteria are separated and the insulin is purified.

In this process, the plasmids are referred to as **vectors** and the process of insertion of the gene into the plasmid as **gene-splicing**.

The advantages of using this method of insulin production are:

• No animals are hurt during the process.

• A large amount of insulin can be harvested.

• The insulin is human insulin and not the insulin of another mammal.

• It is a relatively quick process.

• Bacteria are very small, take up relatively little space and are easy to culture.

Progress check 21.2.

1   Name the enzymes used in producing human insulin by a process of gene transfer. In each case state the use of the enzyme you mention.

2   What term is used for the bacterial plasmids used in genetic engineering?

Genetic modification

Advantages

The production of genetically modified ('GM') crops is controversial. The insertion of genes that make a crop plant resistant to herbicides used to kill weeds that would compete with the crop, is common with crop plants such as soya. Maize has been genetically modified to make it resistant to insect pests and root worm. Research continues to create virus-resistant strains of maize. In these ways, genetic engineering avoids pollution of the environment with insecticides, other pesticides and herbicides. A particular type of rice ('golden rice') has been produced that can make the same chemical (beta-carotene) that causes carrots to be orange. When eaten, the human body converts beta-carotene into vitamin A. All these are clear advantages, but there are disadvantages as well.

Disadvantages

• Herbicides and insecticides are sprayed once, at the time of year best suited to protect the crop from the pests. These pests may already have bred, so spraying again next year is still likely to kill their offspring. With GM crops, the effect on the pests is continuous and thus natural selection is likely to produce resistant varieties of pests much more quickly.

• Genetic modification can have unforeseen side effects on the crop plants. In soya, splitting stems has had the effect of considerably reducing yield.

• GM crops produce pollen containing modified genes. This pollen is blown to other competing plants that, in some cases, have been known to be affected and take on the resistant properties of the crop.

- There is a widespread belief that the consumption of chemically modified foods may have harmful effects on those that eat them.

- There is concern that wealthy multinational companies may achieve a monopoly in the sale of expensive GM crops that cannot be afforded in developing countries of the world.

Chapter summary

■ You have learnt how yeast and enzymes are used commercially.

■ You have learnt how genetic engineering is used in insulin production.

■ Students following the supplementary course have learnt about points for and against the genetic modification of food crops.

Exam-style questions

1 Describe the role of microorganisms during the production of biofuels. [7]

2 Figure 21.4 shows the equipment used in the manufacture of the antibiotic penicillin.

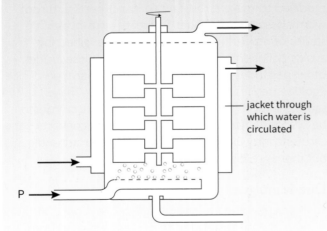

jacket through which water is circulated

P →

Figure 21.4

a i) Explain the purpose of the water jacket labelled in Figure 21.3. [2]

   ii) State what enters the equipment at P and explain its importance. [2]

b Explain the use of enzymes during the production of insulin using genetic engineering. [8]

[Total 12]

3 Explain how a biological washing powder might be better at removing blood stains than a non-biological washing powder. [9]

Human influences on ecosystems

22.01 Food production

Modern technology applied to food production has already greatly increased the amount and quality of food that is produced.

In some countries, the hedges and fences that once divided the land into small units (fields) have been removed so that improved **agricultural** machinery can plant and then harvest crops grown on very large areas of land. This has improved the efficiency and cost of food production.

The size and quality of yield from the crops have been improved by the use of **fertilisers**. Fertilisers can be tailored to supply the correct mineral ions, in the correct proportions for the specific crop being grown.

Insect pests decrease yield by either damaging plant organs such as leaves while the crop is growing, or feeding directly on the part of the crop plant that is of commercial value (often the fruits or seeds). This can be prevented by spraying the crop with **insecticides**. Crops with better quality and higher yields are the result.

Weeds compete with crop plants for light and for mineral nutrition. They can be killed by spraying with **herbicides**.

Selective breeding (see 'Artificial selection' in Chapter 19) is used to produce crops or livestock with greater resistance to disease and a higher yield of better quality produce.

Problems associated with monoculture and intensive livestock production

There is always a danger of disease affecting a particular crop that is grown on a large area of agricultural land, year after year (**monoculture**). The entire crop may fail or have greatly reduced yield (the mineral ions required for that particular crop are used up). This will have a negative effect on the income from that crop as well as on the production of food required to support the community. Sometimes spores of fungal diseases will remain in the soil to affect the crop in following years.

Some crops, such as peas and beans, help to put nitrates back in the soil. However, growing the same crop for several successive years does not help to replenish the mineral ions in the soil; unfortunately, it will deplete the soil of the very minerals that the crop specifically requires.

Intensive livestock production, sometimes called 'factory farming', where animals are kept at high density in confined conditions, produces large quantities of meat, milk and eggs. However, the same problems of disease affecting, this time, large numbers of the animals, can seriously reduce yields. Many consider that such practices are unethical and unacceptable on the grounds of animal rights.

22.02 Food supply

Famine (a lack of adequate amounts of food to support the population) may result from:

- Poverty
- Overpopulation
- Drought
- Flooding
- Crop failure due to disease
- Poor farming techniques
- Unequal distribution of food
- War/political instability
- Or any combination of these.

The world as a whole produces sufficient food to sustain its current population, but some areas greatly overproduce and some areas do not produce anything like enough food to sustain their population. The cost of transporting the food to where it is needed is too high for those requiring it. A 'world solution' needs to be sought.

Overpopulation may be overcome with education on, and availability of birth control methods. Education may also help to improve farming techniques and to help solve some of the problems of flooding where farming techniques are to blame.

Biological research is developing disease resistant crops and improved methods of disease control. An understanding of the effects of deforestation may go some way to improving the problems caused by drought.

22.03 Habitat destruction

The very rapid increase in the size of the human population has already led to the need to increase food production and to find room for the extra housing required. Land once used for **crops** and **livestock** disappears under the increasingly large settlements. Not only does this land have to be replaced, but needs to be even larger in area to supply the larger population. For many years now, the solution has been seen to lie in clearing woodland and using it for food production.

Larger populations require a greater supply of commodities — for example, cars, refrigerators, furniture, bricks and tiles. All these are manufactured from **natural resources extracted** from the soil. The result of that extraction is a depleted and greatly changed natural landscape.

Factories involved in the production of commodities have to dispose of their waste products, as do communities.

Household detergents are discharged into rivers along with sewage. These encourage the growth of small water organisms (algae).

Industrial wastes, such as those that contain mercury (e.g. from paper mills) and copper (heavy metals), are highly toxic to all organisms. It is expensive to remove these wastes in a completely safe way and thus, they too, may be discharged into rivers.

Polluted rivers discharge into seas causing **marine pollution**. Polluted seas lead to contamination of producers in the sea's food chains.

One small fish consumes many smaller contaminated food organisms. One large fish eats many smaller fish and in this way, the amount of poison gradually increases in the organisms along the food chain. When contaminated fish is used as food for humans, harmful levels of poison can be consumed.

In the 1950s and 1960s, mercury waste was discharged into Minamata bay in Japan. It resulted in marine life, particularly fish, being contaminated. The fish were eaten locally and around 10 000 people were affected, many adults suffered from paralysis and many babies from birth defects.

The effect of these activities is to destroy natural habitats. Some species within food chains and food webs that operate within these habitats may begin to decrease in numbers, or disappear completely. The removal of only one link in a food chain can have a major 'knock-on' effect on all organisms within that particular ecosystem.

The undesirable effects of deforestation

When extensive areas are cleared of forests so that the land can be used for agriculture (the process is called **deforestation**), it follows that the habitat of a large number of species is destroyed. Any species living in that one habitat alone face the danger of **extinction**. The loss of trees exposes the soil to wind and rain, allowing the **soil to be lost** as it is washed into rivers.

Once in the rivers, it falls to the bottom as silt causing the rivers then to **flood** the surrounding 'reclaimed' land. As mentioned in Chapter 6 on nutrition, plants take in carbon dioxide for photosynthesis. Forests of trees take in large amounts of carbon dioxide, but when forests are cleared, the carbon dioxide builds up in the atmosphere. Carbon dioxide is described as a 'greenhouse gas', contributing to increased atmospheric temperatures, which can have a global effect.

Progress check 22.1

1 What are the problems associated with 'factory farming'?

2 Make a list of the ways that habitats are being destroyed.

Deforestation has far-reaching effects on the environment.

1 **The loss of soil:**

a) The loss of **humus** in the soil: Leaves from the trees fall to the ground where they decompose forming humus in the soil. Humus provides a steady supply of ions, acts as a sponge, soaking up and holding water in the soil and helps to bind the soil together thus preventing soil erosion.

b) The loss of protection from excessive sun, wind and rain: Trees form a canopy that shields more delicate organisms from the Sun's rays. The canopy also protects the soil from the force of tropical rainfall, and protects the soil, as well as smaller plants and animals from the full force of high winds. Tree roots also help to bind the soil. Removal of the trees therefore leads to soil erosion caused by wind and water.

2 **The effect on climate:** Trees supply enormous quantities of water vapour to the atmosphere as a result of transpiration. Transpiration leads to the formation of clouds. Clouds are carried by the prevailing winds and eventually produce rain, usually in an area some distance removed from that of the trees that released the vapour. Deforestation therefore can lead to distant regions receiving reduced rainfall. At its most extreme, relatively fertile areas can become deserts.

TIP Remember that prevailing winds will normally mean that the climate of the area in which deforestation is occurring may not be the area in which there is most climate change. Deforestation can thus lead to climate change many hundreds of miles away from the deforestation site.

On a global scale, deforestation can reduce the amount of carbon dioxide taken in for photosynthesis. The levels of carbon dioxide in the atmosphere rise, acting as a 'thermal blanket' over the planet, preventing the natural escape of heat from our atmosphere (the 'greenhouse effect'). Much of the solar radiation that is reflected from the Earth's surface is in the form of **infrared radiation**. Greenhouse gases absorb infrared, and radiate it back to the Earth. This is believed to lead to global warming, which may effect the distribution of plants and animals (and eventually melt the ice caps).

3 **The effect on local human populations:** Deforestation is usually prompted by financial motives. Those who benefit are often residents of countries other than those in which the deforestation is occurring. Many local residents are losing their homes and seeing their culture destroyed along with the trees. Many find it difficult to adapt to life-styles that are geared to commercial success. These problems are being faced by those people living in the Amazonian rainforest of South America, which is one of the world's regions of greatest deforestation.

4 **The loss of habitats and of species:** Many consider that humans have an obligation to other species to ensure that they have a suitable habitat in which to live and thus avoid extinction, but there are other, more personal reasons for preventing the extinction of some plant species. Many drugs have their origins in trees of the forest. There may be many more yet to discover, something that can never happen if we allow the extinction of the possible sources of such medications.

22.04 Pollution

Humans are responsible for contaminating the environment with discarded waste and unwanted, toxic (poisonous) chemicals. These are released onto the land or into water (rivers, lakes and the sea). Sometimes they are released into the air, but these pollutants are later deposited on land or in water when it rains.

Common pollutants in this category are as follows:

1   Those used to kill insects: **insecticides**.

    Insects cause considerable harm to the economy, either as pests on crops or as vectors of disease. They can be killed very effectively by insecticides, but insecticides may pollute the environment with the following harmful effects:

    • Useful insects, such as those needed for pollination, may be killed as well.

    • If the livers of animals are unable to break down the insecticide, it may be passed from animal to animal along food chains. Since animals at trophic level A eat many animals from trophic level B, and animals at trophic level B eat many animals from trophic level C (which may have been insects affected by insecticide), animal A receives a very high, and perhaps very harmful level, of insecticide. Some birds are known to have been made sterile by insecticides in this way.

    • They are washed into rivers, killing insects that occupy important places in food webs in rivers. They may also affect insects in lakes and seas.

2   **Herbicides** are sprayed to kill plants that would compete with crop plants for water, light and mineral ions. These, too, can be washed into rivers, thus killing many of the plants that grow naturally in waterways and that form an essential part of food chains and food webs in the water. Also, 'spray drift' may blow onto areas where it kills plants that are no threat to the crop and also form an important part of natural, local food webs.

    Both insecticides and herbicides can be hazardous to the workforce that handles them.

3   **Nuclear fall-out**

    Background radiation from rocks and from the Sun is a form of **ionising radiation** (including alpha, beta and gamma rays) that does not usually reach levels that are dangerous to humans. However, ionising radiation above those levels can cause serious **damage to our DNA** and this can lead to **birth defects** and to **cancer**.

    Nuclear power stations are designed to keep their levels of nuclear **fall-out** well within the safety limits but when accidents occur, the fallout can

be dangerously high. Also, nuclear waste must be disposed of. A problem arises because radioactive materials remain radioactive for many years and once radioactivity affects food organisms (such as grass), it passes along the food chain and can cause **serious gene mutations**.

Other forms of water pollution

We have seen here how chemical waste can affect water courses (see 'Habitat destruction'), but there are other undesirable causes of water pollution.

Since a river would appear to be able to 'carry rubbish away' it would seem to some to be an easy way of getting rid of rubbish. However, plastic bags and bottles can block drains and, together with natural materials such as leaves and branches, help to form dams causing localised flooding. Mammals, birds and fish can mistake pieces of plastic for food and this can causes blockage of the alimentary canal and death.

Harm on a larger scale is caused by releasing **untreated sewage** and **fertilisers** into rivers.

The pollution of water by sewage and fertilisers: large human settlements create a considerable amount of sewage. Tipping sewage directly into streams and rivers can have the following harmful effects:

• Sewage contains pathogenic organisms. If the water is used for human consumption, then diseases, such as cholera, spread.

• Sewage releases nutrient ions as it decomposes. These ions have a similar effect to nutrient ions in any fertilisers washed into rivers. They encourage the rapid growth of water plants (eutrophication). Increased growth leads to increased respiration (mostly as they are decomposed), which uses up so much oxygen in the water that no other life, especially fish, can survive.

Eutrophication and its effects

Within the ecosystem of a **river**, **pond** or **lake**, there will always be producers (plants or photosynthesising protoctists, such as algae) at the base of the food web. The growth of these producers is controlled or limited by the availability of nitrates and other necessary ions within the water. Any action that increases their concentration will encourage the growth of the producers. When fertilisers – particularly nitrogen-based fertilisers such as **nitrates**, ammonium salts and phosphates – are applied to the land, rain will often carry them **into nearby bodies of water**. As a result the number, and possibly size, of the producers increases accordingly. This process is called **eutrophication**.

Eventually, leaves drop off the plants, die and begin to decompose, or the producers themselves die and decompose. This **decomposition** is brought about by **bacteria**. These bacteria are already living in the ecosystem, but they too now increase dramatically in number. Their combined rate of **aerobic respiration** is such that it uses up oxygen faster than it can dissolve from the atmosphere and diffuse throughout the body of water. Organisms that have a high oxygen demand and rely on the dissolved oxygen in the water, such as fish, die. Eventually, only **anaerobic organisms** (e.g. some bacteria) remain alive and continue to live on the decomposing organic matter. This largely lifeless condition of the river or lake is the result of **eutrophication** and does not form part of the process.

TIP

Eutrophication is a reference to the enhanced feeding and growth of aquatic plants. It does not include the subsequent death and bacterial decomposition of these plants and the depletion of oxygen as it is used up by the bacteria of decomposition. All these processes occur after eutrophication has taken place.

The effects of non-biodegradable plastics in the environment

Most plastics have a high resistance to the bacteria that cause decay. Thus, when discarded, they take many years to decompose. We see this from the plastic bags and bottles that disfigure the countryside.

Worked example

a Explain what is meant by eutrophication.

b i) State the causes of eutrophication, and

   ii) explain its effect on organisms living in the habitats in which it occurs.

Answer

a The word eutrophication means 'healthy feeding'. It is normally applied to plant life, and particularly to small green organisms called algae, which live in bodies of water such as ponds, lakes, rivers and also the sea. Healthy feeding leads to excessive growth and large numbers of plants and algae (sometimes called an 'algal bloom') in the water.

   (Note that eutrophication relates only to the excessive growth of the producers, and not to other organisms living in the same habitat.)

b i) The causes of eutrophication can be divided into two – agricultural fertilisers and sewage. It is difficult to judge exactly how much fertiliser to apply to land and impossible to know how much rainfall will follow the application and when it will fall. Fertilisers are readily soluble and thus there is always a danger that they will drain into bodies of water such as rivers and lakes. When this happens, the fertilisers affect the producers in the water in the same way as they are intended to affect the crop plants for which they were intended. The water plants and algae grow especially well. Sewage also contains mineral ions that encourage growth. Sewage is a natural fertiliser and it therefore has a similar effect to that of the artificial fertilisers.

   ii) The effect can be considered both in the short term and in the longer term. In the short term, an algal bloom can completely cover the surface of a river, lake or pond. This can prevent light penetrating to plants that grow beneath the surface, preventing them from photosynthesising and thus causing their death. In the longer term, all the producers will die, and bacteria living in the water (many of which may have been introduced in the sewage) then begin to decompose the dead organic matter. These bacteria are largely aerobic, and thus create great demand on the oxygen content of the water. So great, that they use it all up and there is not sufficient for the respiratory needs of consumers in the habitat such as insects and fish, which also die as a result. The body of water is then unable to support any life. Note, again, that this situation is the result of eutrophication, not part of it.

Effect on aquatic systems

We have seen that plastic waste can harm creatures living in or on water. Water is also vital for drinking and plastic that finds its way into drinking water reservoirs or rivers from which drinking water is extracted slowly releases toxic chemicals over a long period of time. One in particular (bisphenol A) damages the reproductive systems of animals.

Disposal of plastic in oceans affects both marine animals and seabirds. As the years pass, plastic apart from releasing toxins, also breaks down into small pieces, which animals mistake for food and consume. The animal may die, but its body might decompose, releasing the plastic again to pose a threat to other marine animals. Marine species can also at times get tangled in the plastic layer, formed from dumping plastic wastes in the water.

Effect on terrestrial ecosystems

Plastic, being a light-weight material, can be blown considerable distances by the wind. Apart from the unsightly nature of waste plastics, animals may consume plastic that eventually leads to their death – either due to a blocked alimentary canal or respiratory system. Also during the rains, the plastic fallen on roads may be washed into nearby water reservoirs or flow into drains. In many, mostly tropical regions of the world, mosquitoes carry disease organisms – such as those that cause malaria. Mosquitoes breed in stagnant bodies of water, which may well be the result of drains blocked by discarded plastics.

Air pollution

This is caused by the release into the air of gases in larger quantities than would be the case from natural cycles. They are often released from factory chimneys. The two most significant are:

1 **Carbon dioxide:** (see 'Deforestation') Apart from being released by respiration, carbon dioxide is also released during the burning of fossil fuels, thus factories and motor vehicles add to the problem of global warming.

2 **Methane:** Carbon dioxide is not the only greenhouse gas. One that is released into the atmosphere from the decay or organic matter is methane. It is released from bogs and from the alimentary canals of animals. Thus the intensive farming of livestock produces a high yield of methane. A sheep releases 8 kg per year from its digestive system. A human makes the modest contribution of 0.12 kg per year!

Methane is also produced in large quantities by decomposing organic matter – particularly in landfill waste sites.

Acid rain

Sulfur dioxide is a gas that is released whenever fossil fuels are burnt. In industrial areas, the amounts of sulfur dioxide released into the air can be high. It has a harmful effect as a gas since it is linked with bronchitis and heart disease.

In the air, sulfur dioxide dissolves in rain to fall to earth as a dilute solution of sulfuric acid. This is known as **acid rain**, which has the following effects:

- It kills the leaves of some species of plant (e.g. wheat).

- It makes the water of lakes acidic. This acidic water now dissolves toxic chemicals (such as salts of aluminium) present in the mud of the lake and that are insoluble in neutral or alkaline solutions. Fish, for example, are killed by aluminium.

As well as sulfur dioxide, **oxides of nitrogen** are released during the combustion of fossil fuels and, they too, dissolve in rainwater to form **nitrous** and **nitric acid**, adding to the sulfur dioxide in acid rain. In some areas, large areas of woodland have been destroyed by acid rain and in some lakes, the high pH levels have left them devoid of almost all life. It is important, therefore, to look for ways to reduce the emissions of these toxic gases.

- Factory chimneys can be fitted with '**scrubbers**' that spray the gases produced with water droplets. These dissolved the sulfur dioxide before it is released into the air, and the sulfuric acid thus obtained can be used commercially. Scrubbers also remove 'particulate' matter such as soot and dust particles.

- The use of **solar energy** and **wind power** instead of fossil fuels would greatly reduce the release of sulfur dioxide into the air from power stations

- The fitting of **catalytic converters** to motor exhaust systems reduces the emissions of oxides of nitrogen.

Further detail on the greenhouse effect

The solar radiation that arrives at the Earth's surface is in the form of **short-wavelength radiation**. Short wavelength radiation is able to pass through the blanket of greenhouse gases surrounding the Earth's atmosphere. The radiation is absorbed by the Earth's surface but when it is released again from the Earth's surface, it is in the form of long-wavelength radiation, not all of which passes away through the blanket of gases. It is reflected back to the Earth and hence, heat is trapped in the Earth's atmosphere. Too thick a blanket of greenhouse gases means that less of the reflected heat escapes and the Earth's atmospheric temperature rises.

Another example of pollution: the negative effects of the contraceptive pill on water courses

The hormones present in the female contraceptive pill are difficult to remove during the purification of drinking water. The result is that their concentration is gradually increasing. Drinking water that contains an increasingly high amount of female hormone is thus being consumed by the population. At the same time, the average number of healthy sperms produced by otherwise healthy males (their 'sperm count') is decreasing. This has been blamed on the levels of female hormones in drinking water. The theory is supported by research that suggests that increasing numbers of male, aquatic animals, such as fish, are losing their male characteristics and are becoming 'feminised'.

Progress check 22.2

1   You could calculate the contribution made to the greenhouse gases by a flock of sheep – or even a school of children.

2   Make a list of all the ways in which humans are causing pollution to the planet.

22.05 Conservation

Sustainable resources

A **sustainable resource** is defined as **one which is produced as rapidly as it is removed from the environment so that it does not run out.**

Non-renewable resources are therefore very precious commodities and need to be used as economically as possible. Such resources include fossil fuels. The global economy depends fundamentally on fossil fuels (coal, oil and gas) and, until an alternative is developed, measures have to be taken not to waste them.

It is now realised that because trees take a long time to reach maturity, chopping them down for wood products such furniture and paper must not outstrip the rate at which they can be replanted and reach a size suitable for harvesting. Also, if fish are caught and consumed at a faster rate than they can breed, then fish stocks will become seriously depleted.

Recycling of resources

Pressure is taken off endangered species if countries encourage a policy of recycling. Many cars and drinks cans are now made of recycled **metal**. Bottles are made from recycled **glass**.

The need for disfiguring the environment during mining activity is thus reduced. The burning of fossil fuels is reduced since heat is used to extract metals from their ores.

Paper is also recycled, reducing the number of trees that need to be cut down.

Litter is also reduced by recycling, making our environment a more pleasant place in which to live, but an ever-increasing component of everyday **litter is plastic**. Most plastics are **non-biodegradable**, which means that, though they may well be made of organic chemicals, they are not decomposed by bacteria or fungi. It is therefore important that they are recycled whenever possible. The dyes in plastics are difficult to remove and not all plastics are easily recyclable – especially those that have been recycled once already.

Another problem is the very high temperatures necessary to melt the plastic – as this tends to use up valuable fossil fuel resources.

Reuse of bottles

Glass and some plastic bottles are widely reused to conserve resources. If the bottles are used for drinks, sterilisation between uses is an important consideration.

Recycling of drinking water

With the ever-increasing human population, there is an ever-increasing need for safe drinkable (**potable**) water. Some parts of the world are constantly short of suitable water supplies. Water is costly to purify and thus it is important to conserve water wherever possible.

When sewage is properly treated, the water it contains can be recycled. Sewage provides an effective fertiliser and the considerable amounts of water used to carry away the sewage, and that would otherwise be wasted, can be purified even to the extent that it can be returned to drinking water supplies! Figure 22.1 shows the stages in treatment of sewage.

The conservation of species

The need for conservation

We have seen in the section on food webs how the alteration of the natural balance of organisms in a habitat, such as when a **new species** is introduced, can reduce species numbers and even bring about their **extinction**. But there are other several different reasons why the numbers of any species may decline to the level at which it is regarded as **endangered**. **Climate change** can have a marked effect on species numbers. The change in climate could be as a result of deforestation or global warming. Species may be adversely affected by any of the methods of **pollution** mentioned previously, and a number of species of animal have suffered as a result of being hunted to extinction. There is great concern that the numbers of rhinoceros have declined considerably as a result of hunting them for their horns and the numbers of elephant have also declined sharply as a result of the ivory trade.

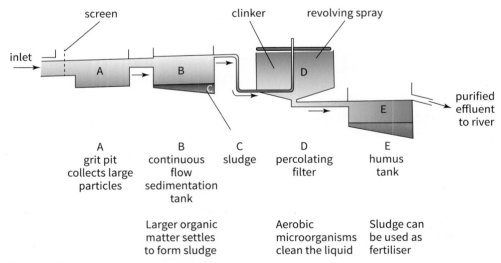

Figure 22.1 How pure water is obtained from sewage treatment

How conservation can be achieved

It is important to identify species whose numbers are declining so that their decline can be halted and reversed before they become endangered.

- **Monitoring** numbers is therefore a crucial part of conservation.

- **Protecting habitats** is essential, but increasingly difficult as more and more land is swallowed up for agriculture and for transport routes. Zoos are not always popular, as some people think that the animals are better in their natural habitats, but many zoos are engaged in **captive breeding programmes** with the aim of eventually re-introducing the animals that have been bred back into the wild.

- **Seed banks** serve a similar purpose for plant species that are endangered, so that they can be grown, their seed harvested and replanted in their natural habitats.

- Most important, too, is the need to **educate** the public to have a respect for other species and an understanding of their needs.

Sustainable development

The increasing human population and the gradual spreading of human settlements to swallow up natural habitats must be balanced against the needs of other organisms on the planet, as well as against the requirement not to deplete vital resources.

This is calls for a policy of **sustainable development** – a term that is defined as **development providing for the needs of an increasing human population without harming the environment**.

Mentioned previously is the problem of depletion of fish stocks. This is being addressed by educating fisherman in the life histories of the fish, by not catching fish of less than a certain size, otherwise they will never reach maturity and have the chance to breed, by **legally imposing quotas** on the number or mass of fish allowed to be caught and by fish farms that follow breeding programmes allowing the release of young fish back into the sea (**restocking**) to breed and maintain numbers.

These strategies have to be internationally agreed, and enforced, so that there is cooperation by all involved. There will always be a temptation to increase income by ignoring the quotas, but that is a short-sighted policy as, eventually, it will lead to fish stocks so low that there is neither food for the population nor income for the fisherman.

A similar argument can be used for maintaining wood supplies from forests and must be implemented if the current trends are to be reversed.

Problems experienced by threatened species

The fewer organisms that there are to breed, then the smaller the variety there will be amongst the parents and therefore, also amongst their offspring. This has two important effects: (1) there may be no members of the population capable of withstanding a change in the environment; (2) there will be a greater chance of genetic weaknesses appearing in the offspring.

The importance of conservation programmes

In order to maintain the wide variety (biodiversity) of living species on the planet, it is important to identify threatened species so that their needs can be addressed before it is too late. Such work is carried out by WWF (the Worldwide Fund for Nature) and CITES (the Convention on International Trade in Endangered Species). Countries throughout the world must then support such organisations with legislation.

All schemes to conserve species are long-term ones. Although it may be difficult to (re)produce the exact environment necessary for the best chances of survival for a threatened species, the aim will always be to prevent extinction or, at the very least, to reduce its likelihood. It is therefore **important to identify vulnerable habitats** as early as possible, as this will enable conservationists to take the necessary steps to maintain species numbers. In doing so, they will also be able to protect the resources that ecosystems provide that go beyond just saving a species from extinction. These include:

- Foods naturally supplied within the ecosystem (e.g. fruits).

- Drugs that may be produced by plants in the ecosystem (e.g. from tree bark – such as quinine).

- Natural fuels from trees.

- The variety of genes necessary to produce variation within vigorous populations of different species.

Chapter summary

- ☐ You have learnt how demands of world food supply may be met, but also about the problems associated with satisfying those demands – including deforestation and soil loss.

- ☐ You have learnt about different forms of pollution – their causes and harmful effects.

- ☐ You have learnt of our responses to pollution through the recycling of resources.

- ☐ You have learnt about the importance of conservation.

Exam-style questions

1 State how insecticides, chemical waste and untreated sewage can pollute bodies of water such as rivers and lakes. [10]

2 Table 22.1 shows the concentrations in the atmosphere over the last 400 years of carbon dioxide and of methane.

year	concentration in the atmosphere/ parts per million	
	carbon dioxide	methane
1600	280	0.75
1800	280	0.75
2000	360	1.75

Table 22.1

a i) State the term used for these two gases when describing their effect on the environment. [1]

ii) Explain what are considered to be their undesirable effects on the atmosphere.

[4]

b Explain the changes in atmospheric concentration of these gases over the last 400 years. [6]

[Total 11]

3 The diagram shows the proportions of different pollutants released into the air in a particular country.

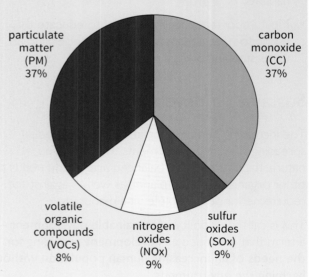

Primary air pollutants

particulate matter (PM) 37%

carbon monoxide (CC) 37%

volatile organic compounds (VOCs) 8%

nitrogen oxides (NOx) 9%

sulfur oxides (SOx) 9%

Figure 22.2

a Explain the dangers to organisms of pollution of the air by i) carbon monoxide and ii) sulfur oxides. [8]

b i) Suggest a pollutant, other than carbon dioxide, that is one of the VOCs mentioned in the diagram. [1]

ii) Suggest and explain a source of the nitrogen oxides. [2]

[Total 11]

Progress check answers

Chapter 1

Progress check 1.1

1 4

2 movement, respiration, sensitivity, growth, reproduction, excretion, nutrition

3 C

Progress check 1.2

1 chitin

2 B

Progress check 1.3

1 1024

2 They join amino acids together to make proteins

Progress check 1.4

1 B

2 B

3 D

Progress check 1.5

There are many possible answers, such as

1. Shape with 3 sides . . . . go to 2

  Shape with more or fewer than 3 sides . . . . go to 3

2. Shape with one curved side . . . . E

  Shape with no curved sides . . . . go to 4

3. Shape containing right-angled corners . . . . . go to 5

  Shape with no corners . . . . D

4. Shape with sides of equal length . . . . A

  Shape with one side shorter than the rest C

5. Shape with all sides straight . . . . B

  Shape with one curved side . . . . F

Chapter 2

Progress check 2.1

1 B

2 Chloroplasts contain chlorophyll for photosynthesis. Roots are in the ground, in the dark and do not photosynthesise, so they have no need for chloroplasts.

Progress check 2.2

1 They are the cell's centre of energy release during respiration.

2 cell wall, cellulose, large central vacuole, cell sap, chlorophyll in chloroplasts

3 20 µm

Progress check 2.3

1 D

2 There are large numbers of both and they are very small, this provides a large surface are for absorption. They both absorb oxygen.

3 Palisade cells as they are nearest to the upper surface of the leaf and thus receive most sunlight for photosynthesis.

Progress check 2.4

1 D

2 Check with your teacher

Chapter 3

Progress check 3.1

1 Check with your teacher

2 Random motion down a concentration gradient

Progress check 3.2

1 The cell membrane.

2 Osmosis is a term applied only for the diffusion of
   water molecules. Osmosis is diffusion because the
   molecules move from a high to a low concentration.
   Osmosis always occurs through a partially
   permeable membrane.

Progress check 3.3

1 The animal cell does not possess a cell wall which
   can resist the pressure of the water entering the
   plane cell. The cell membrane has no mechanical
   strength.

2 C

3 Firm, open leaves are the result of (turgor) pressure
   in the leaf cells. This is caused by water that has
   entered by osmosis pushing against the cell walls.
   When cells plasmolyse, they are short of water,
   there is no longer turgor pressure within the cells
   and the leaf is no longer held flat and open – it
   droops.

Chapter 4

Progress check 4.1

1 i) CHO, ii) CHO, iii) CHON

2 i) fatty acids and glycerol, ii) glucose, iii) amino acids

Progress check 4.2

1 i) iodine solution ii) DCPIP

2 Biuret solution – protein is present

3 Benedict's solution tests only for reducing sugars.
   There are reducing sugars other than glucose (e.g.
   maltose).

Chapter 5
Progress check 5.1

1 Substrate.

2 Enzymes are not used up. After catalysing a reaction,
   they are immediately available to catalyse a similar
   reaction.

3 Those that work in plants, in fungi and bacteria,
   those in organisms living in hot springs, some used in
   washing powders.

Progress check 5.2

1 It will indicate the pH at which amylase works best –
   around pH 7 as found in the mouth and duodenum
   (where it might be a little above pH 7).

2 The substrate molecule.

Chapter 6

Progress check 6.1

1 Carbon dioxide and water are used to make glucose
   with oxygen as a waste product. Light energy is
   harnessed (or trapped) by chlorophyll to form
   chemical energy within the glucose molecule.

2 Nutrition.

Progress check 6.2

1 To ensure any starch present in the leaves is
   converted to sugars and carried away from the
   leaves. Any starch then present in the leaves must
   have been made during the experiment.

2 Each leaf is both the experiment (the green part)
   and the control (the white part).

3 D

Progress check 6.3

1 B

2 Water travels from the soil, through root hairs and across the cortex of the root to the xylem. It then travels up the xylem in the vascular bundles of the stem, then into the veins of a leaf.

Progress check 6.4

1 C

2 epidermal cells, since chloroplasts would hamper the passage of light to the main photosynthesisng cells, the palisade cells, immediately beneath.

Chapter 7

Progress check 7.1

1 B

2 Starvation – a severe lack of food, malnutrition – a diet lacking in the correct balance of constituents.

3 Kwashiorkor – a protein deficient diet giving rise to a swollen abdomen.

   Marasmus – the result of a diet severely lacking in both protein and carbohydrate and containing insufficient protein for growth and insufficient energy for the body's needs.

Progress check 7.2

1 Chemical – the conversion of insoluble food molecules into soluble ones using enzymes, Mechanical – breaking large pieces of food into smaller ones using teeth and the churning action of the stomach.

2 The enzyme protease, in the presence of hydrochloric acid, digests proteins in the stomach.

Progress check 7.3

1 mouth (cavity), oesophagus, stomach, duodenum, ileum, colon, rectum, anus

2 Its walls have many hair-like projections, villi, and it also has pleated lengthways to provide a large surface area for the absorption of digested food, water, ions and vitamins. It is long.

3 A

Progress check 7.4

1 Enamel

2 Molars and premolars

3 To neutralise the acid released by bacteria that dissolves tooth enamel.

Progress check 7.5

1 See Table 7.4.

2 Fats – because the gall bladder stores bile, which is used to break large globules of fat into much smaller ones (droplets) during emulsification. This proves a larger surface area on which the fat-digesting enzyme lipase can work more quickly.

Chapter 8

Progress check 8.1

1 C

2 Sucrose and amino acids (in solution).

Progress check 8.2

1 The volume of water taken up by the leafy shoot.

2 C

Progress check 8.3

1 Transpiration is the loss of water vapour mainly from the leaves of plants. Translocation is the movement of chemicals around a plant, usually as sucrose and amino acids in the phloem.

2 When a plant is photosynthesising, there is a greater concentration of carbohydrates in the leaves than in other parts of the plant. The carbohydrates (in the form of sucrose) are thus transported from the leaves to the other parts, for respiration and for storage. The leaves are thus the source. When the plant is not photosynthesising and there are then more carbohydrates in the stem and roots of the plant, these parts become the source and carbohydrate is carried from them to the leaves for respiration, and the leaves become the sink.

Chapter 9

Progress check 9.1

1 As it travels from the heart to the gills.

2 A

Progress check 9.2

1 pulmonary vein

2 A

3 The weight of the right leg pushes down on the artery in the leg. Pulses of increased pressure in the artery rhythmically push up the leg causing the foot to 'kick.'

Progress check 9.3

1. A

2. C

3. animal fats, cholesterol and salt in the diet, obesity, smoking, stressful life-style, lack of exercise, inherited genes

Progress check 9.4

1 Those cells in greatest number are red blood cells, the larger cells are: phagocyte (left), lymphocyte (top right), dots are platelets.

The 'holes' are not holes at all, they represent the thinner central regions of the red blood cells, through which light passes much more easily.

2 There are no suitable protease enzymes in the plasma to digest them and they are required as proteins to perform their functions (such as antibodies and the proteins involved in blood clotting – e.g. fibrinogen).

Chapter 10

Progress check 10.1

1 direct contact, through the air, by touching contaminated surfaces, through contaminated food, from animals (including pets)

2 a bacteria-proof skin, blood clotting over cuts, mucus that traps bacteria, hydrochloric acid in the stomach, antibodies from white blood cells, phagocytosis by white blood cells

Chapter 11

Progress check 11.1

1 The only access to the atmosphere is via the mouth/nose. These both lead to the lungs, so therefore air must always enter the lungs when we breathe in.

2 It picks up oxygen that has diffused through the alveolus walls. The oxygen is bonded to the red pigment contained in the cytoplasm of the red cell, and forms oxyhemoglobin.

Progress check 11.2

1 So long as the exercise is gentle, oxygen and glucose can be supplied to the muscle fast enough and in sufficient quantities to meet muscles' energy demands.

2 D

Chapter 12

Progress check 12.1

1 Renal vein

2 Faeces are largely undigested cellulose and lignin (fibre) that has never taken part in a chemical reaction in the body. These chemicals are not therefore excretory products.

Progress check 12.2

1 Cortex

2 B

Chapter 13

Progress check 13.1

1 i) muscles and glands ii) A nerve is a bundle or collection of individual nerve cells, neurones are individual cells.

2 Stimulus

Progress check 13.2

1 Only one neurone makes and releases transmitter substance and only the other neurone at a synapse possesses receptor molecules in its membrane.

2 Involuntary or reflex actions are fast, automatic (not controlled by the brain), are short-lived, often protective and each response is always triggered by the same stimulus. All these points are the reverse of voluntary actions.

Progress check 13.3

1 The object is then not in the centre of your field of vision. Its image then falls not on the fovea, but to the side of it. There is a higher concentration of rods outside the fovea than in it and rods are better at seeing in dim light.

2 The line of travel of the punch exactly coincides with the area covered be the blind spot of the eye on that side. The boxer thus may not see the punch coming until it is too late.

Progress check 13.4

1 C

2 C

Chapter 14

Progress check 14.1

1 Glucagon (or adrenalin)

2 Glycogen

Progress check 14.2

1 Homeostasis

2 sweating, fat insulation, shivering

Progress check 14.3

1 B

2 i) slows it down ii) speeds it up

Chapter 15

Progress check 15.1

1 The bacterial cell membrane is delicate and can be damaged. The cell can burst, especially if the water potential in the bacterial cytoplasm is lower than in the tissues of the person taking the antibiotic. Water will enter the bacterium by osmosis.

2 MRSA is a bacterium that, following mutation, is now no longer destroyed by antibiotics commonly available.

Progress check 15.2

1 The body becomes so accustomed to a drug that increased dosage is necessary in order to produce the same desired effect.

2 C

Chapter 16

Progress check 16.1

1 C

2 There is a lack of variation in the offspring, reducing the rate of evolution. Since all individuals are genetically identical, disease affects them all. Offspring are likely to be found very close to their parents and fellow offspring. This leads to overcrowding and competition for resources such as water, mineral ions and light.

Progress check 16.2

1 A

2 C

Progress check 16.3

1 C

2 × 0.6

Chapter 17

Progress check 17.1

1 D

2 i) testes, ii) ova (or egg cells)

Progress check 17.2

1 B

2 Respiration in the mitochondria supplies the energy necessary to power the lashing flagellum that causes to sperm to swim. Relative to its size, the distance covered by the sperm is very large.

Progress check 17.3

1 Blood is lost during menstruation. Blood contains hemoglobin and hemoglobin contains iron. Thus iron is lost during menstruation, which has to be replaced.

2 Sperms may be released before ejaculation and it requires considerable self-control.

Chapter 18

Progress check 18.1

1 At the moment of fertilisation

2 D

Progress check 18.2

1 D

2 C-U-U-A-G-A

Progress check 18.3

1 D

2 They are unmodified and can be modified to perform any function.

Progress check 18.4

1 I

2 I

3 phenotype

Progress check 18.5

1 Key – let **D** represent the dominant allele and **d** represent the recessive allele.

	heterozygous			homozygous			
				dominant			
Parental genotypes	Dd		×	DD			
Gametes	D	d				D	D

	D	D
D	DD	DD
d	Dd	Dd

Offspring/F$_1$ genotypes: **DD, DD, Dd, Dd**

Genotypes: All show the dominant phenotype.

2 D

3 A

Chapter 19

Progress check 19.1

1 Start with parents who are $Hb^A Hb^S \times Hb^A Hb^S$. Draw a full genetic diagram to show how there will be a 1 in 4 chance of their offspring having sickle cell anaemia.

Parental genotypes		$Hb^A Hb^S$		$\times$		$Hb^A Hb^S$	

Gametes	Hb^A	Hb^S		Hb^A	Hb^S

	Hb^A	Hb^S
Hb^A	$Hb^A Hb^A$	$Hb^A Hb^S$
Hb^S	$Hb^A Hb^S$	**$Hb^S Hb^S$**

Offspring genotypes:	$Hb^A Hb^A$, $Hb^A Hb^S$ $Hb^A Hb^S$	**$Hb^S Hb^S$**
	'normal' blood	sickle cell

1 Record your answer as a ratio – rollers to non-rollers – $x : 1$ – discontinuous variation

2 Check the accuracy of your list with your teacher.

Progress check 19.2

1 Adaptive features suit an organism to its environment increasing its chances of survival.

2 Check with your teacher that the adaptive features you mention are related to the environments in which your chosen organisms live.

Chapter 20

Progress check 20.1

1 D

2 Each food chain you select must start with a producer (a green plant), have at least three organisms and read from left to right with arrows – again left to right – between each organism. Check with your teacher that the food chains you have selected are valid for the habitat you select.

Progress check 20.2

1 Aphids, caterpillars and beetles do not eat fruits – therefore unaffected; flies do, so they decrease in numbers.

2 Depending on what it fed on and how effective it was at obtaining its food compared with other secondary consumers, there are many possible answers. Discuss your suggestions with your teacher.

Progress check 20.3

1 Check the accuracy of your food web with your teacher.

2 D

3 D

Progress check 20.4

1 Lack of food, predation, disease, poisons (e.g. build up of excretory materials), climate (abiotic factors).

2 D

Chapter 21

Progress check 21.1

1 The results show that the enzyme protease in bio-washing powder works better at 40 °C than at any other temperature. 70 °C destroys the enzyme, but this is nearer the temperature at which non-bio powders work.

This shows that the protease works best at a neutral pH (though, if the protease used was stomach protease, it would be the one in acidic conditions that shows the most effective removal of the blood stain).

2 With differing temperature, there would be unlikely to be any difference in the results. At 0 °C, it would need a great deal more than double the time to show any signs at all of digesting the blood. At 70 °C, a denatured enzyme will still not work, even if the time is doubled. With the differing pHs, again, double the time would be unlikely to show any difference in the wrong pH, but it might allow a better result in the correct pH for the enzyme.

3 It is likely to be an enzyme.

Progress check 21.2

1. Restriction enzymes for i) 'cutting' the gene from a human chromosome and ii) 'cutting' open the bacterial plasmid; and ligases for joining the human gene with the bacterial plasmid.

2. Vectors

Chapter 22

Progress check 22.1

1. Disease can affect large numbers of animals at a time. It can have a serious effect on yields of meat or milk. Many people consider it unacceptable to keep large numbers of animals in cramped conditions.

2. Clearing woodland (deforestation), extracting resources from the soil, discharging household detergents, sewage into rivers, as well as heavy metals from industry. Heavy metals, in particular, also contaminate seas. Even the removal of one link in a food chain can have a serious knock-on effect on other organisms.

Progress check 22.2

1. a flock of 50 sheep: $50 \times 8 = 400$ kg per year

   a school of 500 children (and teachers!) $= 500 \times 0.1 = 60$ kg per year

2. Insecticides, herbicides, nuclear fallout, sewage, fertiliser, plastics, carbon dioxide, methane, acid rain, contraceptive pills – perhaps you can add to the list?

Exam-style question answers

For the exam-style questions with allocated marks, answers are provided in the form of a guide to marking your answers. The answers have been presented as a list of statements that could gain marks, but are not the complete answer. When you are checking your own answers, please note:

- *Where answers are given in bullet lists, each bullet point is worth a potential mark.*

- *There may sometimes be more than one way of making the scoring point. These alternatives are shown by an oblique line (/) or by the word OR. However, you can only give yourself one mark for each alternative.*

- *For some answers, there are more points for which you can get a mark than the total number of marks available for the question. For these, any combination of marking points is acceptable – up to that total number of marks.*

Chapter 1

1   1 mark for each correct row of ticks [5]

     1 mark for each correct identification [5]

	1a	1b	2a	2b	3a	3b	4a	4b	5a	5b	name of arthropod
A											
B	✓			✓			✓			✓	Anopheles
C		✓	✓								Ornithodorus
D	✓				✓						Pulex
E	✓			✓	✓						Musca
F	✓			✓			✓	✓			Periplaneta

2   Any two, 2 marks per line, from the following:

   (a) Insects, three pairs of legs/six legs/body has three sections/wings

        Arachnids, four pairs of legs/eight legs

        Crustacea, two pairs of antennae [4]

   (b) Any two, 2 marks per line from the following:

        Out of sight, to avoid predators/being eaten

        It is damp, to avoid drying out

        Avoids the Sun, avoids the effect of UV light

        The presence of suitable food, it feeds on other invertebrates

        It is cooler, to avoid overheating/ drying out [4]

3 (a)
- viruses are the most primitive of organisms
- so primitive that they are incapable of normal life outside a living cell
- they have to invade a living cell of another organism before they can reproduce
- this invasion causes damage to the cell
- an infection of many cells with viruses creates in the host symptoms of disease
- thus many viruses are pathogens (or disease-causing organisms) and are responsible for relatively minor diseases such as influenza as well as major illnesses such as HIV/AIDS.
- their minute size means they are visible only under an electron microscope
- they are not affected by antibiotics so virus diseases difficult to treat
- recovery from a virus infection usually depends on the infected organism's own response, such as antibody production [6]

  (b)
- viruses are very much smaller than bacteria
- less than 300 nm in size
- 50 times smaller than bacteria
- they are surrounded by a protein coat (a capsid)
- whereas bacteria are surrounded by a cell wall – and sometimes a slime layer round the cell wall
- both types of organism contain nucleic acid
- virus may have either DNA or RNA
- bacteria have only DNA [5]

Chapter 2

1 (a) (i) blood [1]

    (ii) Cell A hemoglobin

         Cell B chlorophyll [2]

(b) • chloroplasts

    • cytoplasm

    • air spaces

    • cell wall

    • cellulose

    • cell sap

    • vacuole [7]

2 (a) ✓ against organ, [1]

(b) • kidney + is an organ, ureter + is an organ, bladder + is an organ

    • each contain tissues working together, one constituent tissue mentioned (e.g. muscle tissue or nerve tissue)

    • correct function of any of the named organs (e.g. kidney for filtering blood OR ureter for taking urine from kidney to bladder OR bladder for storing urine)

    • all three of these organs work together + to form an organ system that performs the process of excretion OR makes up the excretory system max [6]

3 • xylem has two functions

• to conduct water and mineral ions from the roots to the aerial parts of the plant (stems, leaves and flowers)

• and also to support these parts, holding them up towards the sunlight and resisting wind forces.

• for conduction, xylem vessels stretch continuously from the roots to all parts that they supply

• they are in the form of narrow tubes placed end-to-end

• they contain no obstructions

• they are held open by the lignin that is deposited in their walls

• for support, lignin is a strong, tough chemical (commonly called wood).

• it is able to support the weight of plants, even those capable of growing to great height

• the xylem forms part of a vascular bundle, and

• vascular bundles are arranged around the stem forming strengthening rods that resist bending

• in a root, they also resist pulling strains that will help to keep the plant anchored in the soil max [10]

Chapter 3

1 • CO_2 moves in during daylight

• O_2 moves out during daylight

• water vapour moves out more in the light

• or more when the stomata are open

• CO_2 moves out in darkness

• O_2 moves in in darkness

• salts/ions move into cells

• water moves from cell to cell max [6]

2 (a) • water enters the tubing

    • increasing its volume OR making it fatter/larger

    • increasing its pressure OR making it firmer [3]

(b) • molecules have energy

    • they have kinetic energy

    • and are always moving

    • from high to low concentration

    • OR down a concentration gradient

    • dye moves out of the tube

    • by diffusion

    • water moves in

    • by osmosis

    • OR water moves in by diffusion

    • because there is a greater concentration of water molecules outside the tube than inside it

    • OR there is a lower concentration of water molecules outside in the water than in the tubing (or solution)

- only water molecules can pass through the partially permeable membrane
- OR sugar molecules cannot pass through the partially permeable membrane

max [7]

3 (a) (i)
- sap in the cell vacuoles of the root hair cells of the plant has a lower water potential than the soil water
- the cell membrane of the root hair cells is partially permeable
- thus osmosis will occur causing the water to enter the cells
- water potential of the cell sap in the cells of the root lying next to the root hairs now has a lower water potential than the root hair cells
- so water moves into them from the root hair cells
- again by osmosis
- this allows more water to enter the root hair cells from the soil
- this is a continuous process

max [3]

(ii)
- if mineral ions are in very short supply in the soil, they are likely to be less concentrated than they are in the plant,
- so energy must be used to draw the ions into the plant *against* a concentration gradient
- this process is called active transport
- the ions are absorbed through the root hair cells
- which are present in very great numbers
- allowing the maximum uptake of ions

max [3]

b)
- waterlogged soil will contain very little if any air
- air supplies oxygen for the root hair cells to respire
- decreased respiration rate leads to a reduced respiration rate
- reduced respiration rate decreases the energy available for active transport

- thus active transport slows down
- slowing the rate of uptake of ions that are in short supply

max [4]

Chapter 4

1 (a)
P complex carbohydrate or starch or glycogen or cellulose

Q proteins or polypeptides

R fat [3]

(b) (i)
- carbon
- hydrogen
- oxygen [3]

(ii) nitrogen [1]

(c)
- identical proteins have identical sequences
- of amino acids
- sequence depends on the sequence of bases found in the DNA
- of genes
- same genes means same sequence

max [5]

2 (a)
- double
- helix [2]

(b)
- in chromosomes
- in the nucleus [2]

(c)
- both of the Gs linked to C
- all three of the As linked to T
- both Ts linked to A
- C linked to G
- i.e. C-T-T-A-A-C-G-T

[4]

3
- starch is a carbohydrate
- that is it is made from linked glucose molecules
- all the glucose molecules are the same (or identical).
- proteins are made from a large number of amino acids
- there are (around) 20 amino-acids.
- protein molecules are very long molecules

- thus there can be a very large number of different sequences of different amino acids in a protein molecule   max [5]

Chapter 5

1 (a) • line showing decrease in rate

       • matching shape of left hand side

       • meeting *x* axis   [3]

   (b) • rate increases

       • to optimum/fastest

       • at around body temperature/37–40 °C,

       • decreases

       • to zero

       • stays at zero when temperature returns
                                                        max [5]

2 • heat supplies energy

   • causing the substrate to move more quickly into the active site of the enzyme **OR** the product leaves the active site more quickly

   • the rate of the reaction speeds up

   • to the optimum temperature **OR** body temperature **OR** 37 °C

   • the enzyme is destroyed/denatured,

   • at temperatures above 60 °C

   • the shape of the active site changes

   • the substrate no longer fits

   • as a key no longer fits a lock

   • the substrate is the key and the enzyme is the lock   max [8]

3 (a) • enzymes are described as 'specific', which means that each enzyme controls only one reaction

       • they possess an active site, which exactly matches the molecular shape of part of the substrate molecule on which they work

       • for the reaction to occur, the substrate molecule must fit exactly into the active site

       • and be held there while the substrate molecule is converted into its product(s)

- during the reaction, the enzyme and substrate molecules are bound together

- in what is called an enzyme-substrate complex, without which the reaction could not occur   max [4]

   (b) • there might be a tendency for product molecules, immediately after being formed during an enzyme-controlled reaction to re-join

       • and thus revert to the substrate molecule

       • this is prevented by chemically changing the newly exposed parts

       • so that they no longer attract one another

       • when water is the molecule that brings about this change, the type of reaction involved is called an hydrolysis   max [3]

Chapter 6

1 (a) (i) • label to a palisade cell (any one of the eight vertical cells just beneath the upper horizontal layer)

           • label to a spongy cell (any one of the large, roughly oval cells – nine to the left of C, and one below and to the right of C)

           • label to a guard cell (there is one either side of D)

           Deduct one mark for each cell, above three in number, that is incorrect.   [3]

       (ii) • carbon dioxide + water

            • → glucose + oxygen

            • light + chlorophyll (shown above or below the arrow in the equation)
                                                        [3]

            $6CO_2 + 6H_2O \rightarrow C_6H_{12}O_6 + 6O_2$, light + chlorophyll (shown above or below the arrow)   [3]

   (b) • destarched plant

       • plant kept in the dark for 8+ hours

       • is placed in light

       • for 8 (or more) hours

       • a leaf is removed

- boiled in water
- boiled in alcohol
- softened in water
- iodine (solution) is added
- blue black colour appears where the leaf was green
- + yellow or brown where the leaf was white
- starch is made only in green parts
- starch is a product of photosynthesis

max [9]

2 (a)
- the leaf is green in the parts that contain chlorophyll
- in chloroplasts
- of mesophyll cells
- light energy
- is converted to chemical energy
- in glucose/starch molecules
- the white parts are the control for the experiment max [4]

(b)
- in the light the guard cells are photosynthesising
- making sugar
- which passes into the vacuole (or cell sap)
- the concentration of the vacuole rises
- drawing water in by osmosis (or diffusion)
- the cell becomes turgid **OR** the pressure in the cell increases
- the outer cell wall stretches more than the inner one
- the outer cell wall is thinner than the inner one **OR** the inner one is thicker than the outer one
- the cell becomes curved/crescent-shaped/banana-shaped
- guard cells lie in pairs
- therefore a gap appears **OR** stoma opens **OR** pore opens between them max [9]

3 (a)
- up to point P it is the carbon dioxide concentration that is limiting the rate of photosynthesis
- since an increase in carbon dioxide concentration increases the rate of photosynthesis [2]

(b)
- temperature is now limiting the rate of photosynthesis
- since increasing temperature increases the rate.
- it cannot now be carbon dioxide concentration since as it increases, the rate of photosynthesis remains constant max [2]

(c)
- light is a third possible limiting factor.
- if it was a limiting factor here, then an increase in light intensity would increase the rate in both graphs in Figure 6.12
- if not, then the graphs would not be affected

(NB A lack of water would have the effect of closing stomata – restricting the entry and thus the availability, of carbon dioxide – which the graph shows is not the case.) max [2]

Chapter 7

1 (a)
- starch
- simple sugars/glucose
- proteins
- protease/pepsin/trypsin
- fats
- lipase [6]

(b) (i)
- they are broken down **OR** deaminated
- urea is produced
- carbohydrate is also produced
- and it is stored as glycogen max [3]

(ii)
- glycogen is changed to glucose
- under the effect of glucagon or adrenaline
- the glucose is released into blood max [2]

2 (a) (i) • from the outer layer **OR** epithelium [1]

     (ii) • absorption

         • of fats **OR** of fatty acids + glycerol [2]

     (iii) • hepatic portal vein [1]

  (b) • respiration

     • occurs in mitochondria

     • energy is released

     • and used for active uptake **OR** active transport

     • of glucose/amino acids

     • the microvilli increase

     • the surface area for uptake max [4]

3 • a ball of food

  • called a bolus

  • is pushed along the alimentary canal by a series of waves of muscular contraction called peristalsis

  • the intestine walls contain a layer of circular muscle

  • surrounded by a layer of longitudinal muscle

  • the circular muscle contracts behind the bolus

  • decreasing the size of the lumen of the alimentary canal

  • while the longitudinal muscle relaxes

  • the wave of contraction of circular muscle behind the bolus pushes the bolus forwards

  • after the bolus has passed

  • the circular muscle relaxes

  • and the longitudinal muscle contracts

  • to open the lumen ready to receive the next bolus max [10]

Chapter 8

1 (a) • root [1]

  (b) • xylem clearly labelled (the star or 'X' in the very centre)

     • phloem clearly labelled (any of the 4 patches shown within the arms of the star) [2]

  (c) • xylem + water,

     • and mineral salts/ions

     • enter the root from soil

     • via root hairs

     • and arrive at the xylem via the cortex max [4]

     • phloem + sucrose/sugar (glucose is not acceptable)

     • and amino acids

     • come from leaves

     • down the stem max [3]

2 (a) • it is carried in xylem

     • of a leaf vein **OR** vascular bundle

     • it enters a mesophyll **OR**

     • it enters a palisade cell **OR**

     • it enters a spongy cell

     • by osmosis **OR** by diffusion

     • it then forms a water film on the cell surface (or on the cell wall)

     • it evaporates into the air spaces (in the leaf)

     • then diffuses out through stomata max [8]

  (b) • transpiration/loss of water **vapour**

     • creates a water potential gradient

     • which pulls up water from the stem

     • (water) molecules

     • are held together by

     • cohesive forces

     • other forces help such as capillarity **OR** root pressure max [5]

3 (a) • when the atmosphere is dry, there is a greater diffusion gradient between the concentration of water vapour in the intercellular spaces in the leaves and in the surrounding atmosphere

     • water vapour thus diffuses out of the leaf at a faster rate

     • as the concentration of water vapour in the atmosphere gradually increases, so the diffusion gradient gradually decreases

- and thus the rate at which water vapour leaves the leaves by transpiration gradually decreases water is also being taken up to be supplied to the leaves for photosynthesis max [3]

(b)
- in the dark, the rate is never as high as the rate in the light since, despite the concentration gradients being that same as in the light
- the loss of water vapour is restricted since light is required to open the stomata.
- in the dark, sugar is not made in the guard cells
- they do not become fully turgid and thus do not open up the stomata between them
- it is rare for stomata to close completely, so a small amount of transpiration occurs
- leaves are not using water for photosynthesis max [4]

Chapter 9

1 (i)
- they may eat animal fats/cholesterol
- that block of blood vessels
- such as the coronary arteries
- they may be overweight
- lead a stressful life
- smoke
- or take insufficient exercise max [5]

(ii)
- inheritance OR heart disease tends to run in families OR a person may inherit genes that make heart disease more likely
- the risk increases with age
- males more susceptible to heart disease than females [3]

2 (a)
- left atrium/i) correctly indicated
- tricuspid valve/ii) correctly indicated
- right ventricle/iii) correctly indicated [3]

(b) (i)
- (inferior/superior) vena(e) cava(e)

(ii)
- (systemic) aorta [2]

(c)
- blood no longer goes to the lungs OR the pulmonary circulation bypassed OR the blood no longer obtains blood from the lungs
- the machine oxygenates the blood
- blood is no longer pumped by left ventricle
- the machine pumps the blood
- to rest of the body [5]

3
- Blood is kept circulating by a muscular pump called the heart
- The heart pumps rhythmically and continuously
- with the stronger ventricles responsible for imparting pressure to the blood – particularly the left ventricle that has to push blood to all parts of the body
- a system of one-way valves in the heart ensures that the blood is always forced in the right direction
- arteries take blood away from the heart and have thick muscular walls
- as a wave of pressure passes, the muscles in the artery walls contracting against the increased pressure
- otherwise the arteries would 'balloon' and blood pressure would be lost
- enough pressure remains to ensure that blood reaches all parts of the body
- blood is returned to the heart now under greatly reduced pressure
- and in most animals having to flow in some areas of the body, against gravity
- thus, to prevent the backflow of blood, veins have valves
- the contraction of any muscles that may lie near a vein will help to push blood in the vein from one set of valves to the next
- this is assisted by veins being thin-walled vessels

Chapter 10

1 (a)
- keep away from infected people OR isolate infected people
- keep away from infected animals

- use anti-bacterials (disinfectants) on food preparation surfaces
- keep raw meat away from other foods
- wash hands before preparing food
- and after visiting the lavatory
- keep food in refrigerators
- sewage/organic waste must be disposed of hygienically   max [6]

(b) (i) • defence
- against pathogens
- by antibodies   [3]

(ii) • in active immunity
- lymphocytes produce antibodies
- in response
- to pathogens **OR** in response to antigens
- which could be introduced through vaccination
- memory cells ensure antibody production
- is faster following further exposure
- passive immunity
- antibodies
- supplied by some other organism such as mother to child in the uterus
- **OR** in her milk   max [5]

2 • when a pathogen, which might be a bacterium or a virus enters the tissues of the body
- lymphocytes manufacture antibodies
- antibodies are proteins that made in response to chemicals called antigens found on the surfaces of the bacteria
- their variable ends are specific to those particular antigens
- and prevent the pathogen from entering cells
- and stop them developing or reproducing by sticking or clumping the pathogens together
- these clumps are then ingested by phagocytes that digest them

- some lymphocytes retain the ability to produce the same antibodies again at very short notice if required
- and are called memory cells   max [8]

Chapter 11

1 (a) • at first the person is breathing normally
- at 5 minutes began exercise
- breathing becomes quicker
- deeper
- stops exercise at 10 minutes
- no immediate change to breathing pattern
- gradually returns to normal
- at 15 minutes   max [7]

(b) (i) • alveoli
- of the lungs   [2]

(ii) • there are many of them
- giving a large surface area
- they are thin walled
- have a layer of moisture
- and a good capillary **OR** blood supply
                                                        max [3]

2 • refer to (the need for) energy release
- first aerobic respiration occurs
- CO_2 is removed by the blood
- then anaerobic respiration occurs
- lactic acid is produced
- it builds up in the muscle
- creating an oxygen debt
- lactic acid continues to be removed after exercise is finished
- lactic acid may cause muscle cramp   max [6]

3 • the external intercostal muscles relax
- allowing the ribs to swing downwards and inwards
- the diaphragm relaxes
- allowing it to dome upwards

- these muscular movements decrease the volume of the thorax (from front to back and from top to bottom, respectively)

- the decrease in volume causes an increase in pressure inside the thorax

- this increase in pressure forces air out of the alveoli

- through the bronchioles, bronchi, trachea and out into the atmosphere through the nose and mouth

- force can be added to this process by contraction of the internal intercostal muscles that contract to pull the ribs down   max [7]

Chapter 12

1
- hot weather/high temperatures

- and exercise

- lead to water loss by sweat

- leading to low volumes of water to be lost in urine

- increasing its concentration

- a large amount of liquid consumed leads to a high urine volume

- with a low concentration

- a diet containing high protein **or** salt

- leads to urine with a high concentration

- of urea or salt

In all cases, the reverse points would be acceptable, e.g.:

low temperatures

inactivity

reduce water loss by sweat   max [8]

2 (a)   urea concentrations are rising/increasing **or** a correct readings from graph + units given (e.g. blood urea concentration stood at 260 mg per dm³ before treatment) kidneys were not (adequately) removing urea from blood

max [2]

(b) (i)   9.6 hours/9 hours 36 minutes

[1]

(ii)   170 mg per dm³   [2]

(c)   protein is digested to amino acids that are absorbed into the blood, the excess is broken down in the liver **OR** excess is deminated in liver, which increases the urea content of blood   max [3]

(d)   the washing fluid/bathing fluid /dialysis fluid contains glucose at the required concentration, thus there is no concentration, gradient and no diffusion of glucose from blood   max [4]

3
- chemical reactions that take place in cells produce chemicals that are useful to the body

- and chemicals that have little or no value

- if the chemical has no value, it must be removed from the body

- one important chemical reaction that takes place in all cells is respiration

- products of the reaction, apart from energy, are water (which is essential in cells and therefore not a true waste product) and carbon dioxide

- carbon dioxide is a waste product

- it is carried from all cells to the lungs in the blood

- and is then breathed out (excreted) from the lungs to the atmosphere

- when we eat protein, we are taking in a number of individual amino acids that are in excess to our body's needs

- after digestion in the alimentary canal, the excess amino acids are absorbed from the ileum

- and taken in the blood to the liver where a chemical reaction converts them into urea

- this waste product is then taken by the blood to the kidneys where it is removed from the blood

- and then from the body (excreted) in a solution called urine   max [10]

Chapter 13

1 (a) receptor **OR** i) labelled (the upper extreme right-hand eye-shaped structure)

   effector **OR** ii) labelled (any one of the six dots on top of the bottom right striped structure),

   relay neurone **OR** iii) labelled (any part of the circle or the line attached beneath it that lie just to the right of the central oval) [3]

   (b) • muscles

   • glands

   (in either order) [2]

   (c) • stimulus (or named stimulus such as heat) is detected by a receptor

   • an impulse

   • travels along a sensory neurone

   • to a synapse* then a relay neurone

   • followed by a motor neurone

   • then to an effector **OR** to a muscle **OR** gland

   • which contracts **OR** releases a chemical

   No penalty for not recognising the effector as a muscle

   * Or the synapse mark could be mentioned between relay and motor neurones

   max [6]

2 • neurotransmitter substance/named substance, e.g. acetylcholine or noradrenaline

   • is released from vesicles

   • into the synapse

   • it diffuses across gap/synapse

   • binds with receptor molecules

   • on membrane of next neurone

   • and creates a new impulse max [7]

3 • the feeling of nervousness is largely because the adrenal glands are releasing the hormone adrenalin

   • in greater quantities than are released during normal everyday life

   • the effect of adrenalin is to increase the heart rate

   • and to cause the person to breathe more deeply

   • it also diverts blood away from organs such as the intestines that will not be needed for maximum muscle power

   • ensuring maximum blood flow to the muscles

   • an athlete about to run a race requires the best conditions for energy release in his or her muscles

   • this demands the most efficient supplies of oxygen and of glucose

   • a faster heart rate ensures that greater quantities of both are being delivered to the muscles

   • and carbon dioxide will be quickly removed

   • the deeper breathing will ensure that the red blood cells are carrying the maximum quantity of oxygen

   • better supplies of glucose and oxygen to the brain will also sharpen the athlete's senses

   max [10]

Chapter 14

1 • temperature changes are detected by receptors – A

   • in the skin

   • and in the brain

   • when body overheats there is increased sweat from sweat glands – B

   • sweat evaporates

   • heat is lost

   • when body temperature falls

   • the body is insulated by fat in skin

   • and by hair (on head)

   • shivering

   • generates heat max [6]

2 *Credit would be given if the answer starts by describing a body temperature below normal. In that case all points shown with * would be the opposite of those here, and, in order, would be: below, below, vasoconstriction, less, less, less, above, more)*

   • temperature is controlled within set limits

   • when the body temperature is above* normal **or** above* the set point

   • vasodilation* occurs

- of the arteries **OR** arterioles
- more* blood passes to capillaries
- more* sweat is released from sweat glands
- more* heat is lost
- when body temperature is below* normal vasoconstriction occurs
- less* heat is lost
- thus any deviation causes a response that returns temperature to the set point max [8]

3
- the tip of the plumule of a seedling is the receptor for the stimulus of light
- and it is also the place where auxins are made
- Seedlings A can therefore neither receive the one-sided light stimulus
- nor produce auxins that stimulate growth.
- seedlings B can both receive the one-sided light stimulus
- and produce auxins
- the auxins are destroyed by or migrate away from the light
- and cause an increase in their concentration on the dark side
- the dark side of the shoot grows faster than the light side
- causing the plumule to grow towards the light coming through the hole in the box
- seedlings A do not grow and do not show a growth curvature towards the hole
- they remain straight and shorter than seedlings B max [10]

Chapter 15

1 (a) (i) bacteria [1]

(ii)
- they have become less effective
- particulary so for ciprofloxacin
- which is now the least effective of the three
- methicillin is the most effective [4]

(iii) methicillin [1]

2
- antibiotics have been overprescribed
- bacteria mutate
- making them resistant
- patients may not have completed their course of antibiotics
- more resistant strains of pathogen survive
- these surviving strains reproduce
- and pass on their resistance to the following generations max [6]

3 (a)
- alcohol slows the reaction time
- making accidents more likely
- which is of particular concern if a person is driving a vehicle or operating machinery
- alcohol is consumed by many to produce a sense of well-being
- but, in excess, can lead to a loss of self-control
- which can involve violence max [5]

(b)
- people can become addicted to alcohol
- this can prove very expensive
- and the person's family may suffer as a result of insufficient income
- long-term consumption of alcohol can be harmful to a person's health
- it can lead to withdrawal symptoms such as shaking, confusion and mood changes
- it can cause liver damage (cirrhosis) which can prove fatal
- the person may have to be admitted to hospital for expensive treatment – taking the places in hospital of people suffering from diseases that are not self-inflicted max [5]

Chapter 16

1 (a)
- B identified as large anthers
- A identified as long filaments
- the anthers hang out of flower
- and are easily blown by wind
- C identified as feathery stigma

- held clear of the flower
- where it can more easily catch pollen in the wind max [5]

(b) Any three from:
- large amounts of pollen
- which is dry/dusty
- light
- drab colour
- or lack of colour in the flower
- nectar or scent would be absent [3]

2
- there will be a greater genetic variation in gametes
- a greater genetic variation in zygotes **OR** in the seeds
- leading to a greater variety of offspring
- since there is a mixture of genes from two plants rather than one
- offspring have greater capacity to respond to environmental change
- cross pollination means less chance of genetic defects max [6]

3
- wind-pollinated plants produce light, dry and dusty pollen
- in very large quantities
- it is light so that it can be easily blown away
- and carried by the wind
- it is dry so that it does not stick to the anthers that made it
- and it is produced in large quantities as only a very low percentage of the pollen grains produced will ever reach the stigma of another flower of the same species
- there is thus a great wastage of unsuccessful pollen grains
- insect-pollinated flowers produce heavier, stickier pollen grains
- in much smaller quantities
- the pollen grain's mass is less important if it is be carried by a relatively large insect such as a bee.
- it needs to be sticky to stick to the bee's body

- and, as the bee will fly directly to another flower, which has a fairly high chance of being of the same species, it is a much more certain method of pollination
- and therefore quantities produced can be low max [10]

Chapter 17

1 (a)
- vasectomy involves cutting
- of the sperm duct (or vas deferens)
- between the testes and bladder
- sperms cannot pass from the testes to the penis
- or to the partner [4]

(b)
- forms a barrier
- between male and female
- during intercourse, sperm cannot penetrate/ do not pass to the partner/cannot meet the ovum
- **named** STI (e.g. HIV/AIDS/syphilis)
- the type of pathogen involved
- virus for HIV/AIDS
- bacterium for syphilis
- there is no physical contact between partners
- the pathogen less likely to be passed
- the condom must be used correctly (in place before intercourse and carefully removed afterwards) otherwise risk is still present max [6]

2 (a) (i)
- growth of hair on face **OR** under arms **OR** above genitalia **OR** on chest **OR**
- larger larynx **OR**
- deeper voice **OR**
- larger genitalia **OR**
- sperm production **OR**
- muscle development

(ii) testosterone

(iii) testis [3]

(b) (i) • starts rising day 11/12

• peaks at day 21/22

• falls to initial level by day 28 [3]

(ii) • it is increasing as ovulation occurs

• as it is responsible for the production

• of the spongy lining of uterus,

• ready for implantation of the fetus

• if fertilisation occurs, it also inhibits or prevents the production of FSH

• and prevents the development of **OR** the release of further ova (egg cells)

max [5]

3 • there are two particular excretory products in a fetus

• they are carbon dioxide that is the excretory product of respiration that occurs in all the cells of the fetus

• and urea, which is manufactured as a result of deamination of excess amino acids in the fetus's liver

• they both pass along the umbilical cord

• to the placenta, both carbon dioxide and urea diffuse from the fetus's capillaries to its mother's blood

• they are then carried along her vena cava to her heart

• from the right atrium and ventricle they are pumped to the lungs

• where carbon dioxide diffuses out of the blood into the mother's alveoli

• to be breathed out, via bronchioles, bronchi and mother's trachea

• blood continues to the left side of the heart

• and into first, the left atrium, then the left ventricle

• the ventricle pumps the blood, still containing urea, along the aorta

• and via the renal artery to a kidney

• here the urea is filtered from the blood through a glomerulus*

• into the bowman's capsule*

• then via the rest of the tubule*

• and the ureter

• it is stored in the mother's bladder until being released as urine through the urethra [10]

Chapter 18

1 • alleles

• gametes

• meiosis

• haploid

• heterozygous

• genotype

• omozygous

• recessive [8]

2 (a) • recessive allele (both words are required) on the X chromosome

• males have one X chromosome

• males have only one allele for colour vision **or** the recessive allele always shows in male [4]

(b) Credit would be given for each row separately. Since upper and lower case letter C are so similar – any letter in upper and lower case could be used but X and Y would not be acceptable.

One mark per row and one for identifying the sex of each offspring.

	(mother) Cc		×	(father) C –	
Gametes	C	c		C	–
Genotypes of offspring	CC	C –		Cc	c –
Phenotypes of offspring	normal colour vision				colour blind (male)

[5]

_____

* points that apply to Supplement-level answers

3　One mark for each correct full line.

	father		mother		
	$I^A I^o$		$I^A I^B$		*(You need to link each parent to their genotype)*
Gametes of parents	I^A	I^o ×	I^A	I^B	*Don't forget to say that this line represents the gametes*

Possible combinations at fertilisation:

	I^A	I^o
I^A	$I^A I^A$	$I^A I^o$
I^B	$I^A I^B$	$I^B I^o$

↓

This is the only possible group B

Therefore, the chances are that 1 in 4 of the offspring will have blood group B (a 25% chance)　　　[8]

Chapter 19

1　continuous:
- there is a range of phenotypes
- between two extremes
- there are many intermediates
- e.g. height/body mass/foot size/skin colour

discontinuous:
- there are few phenotypes
- with no intermediates
- e.g. blood groups/ tongue rolling　　max [6]

2　(advantage) in the heterozygote **OR** person with sickle cell trait:
- it provides resistance to malaria

(disadvantage) in homozygous recessive person:
- the person has abnormal hemoglobin
- misshapen red blood cells
- that get stuck/block capillaries
- cannot carry enough oxygen
- any two other physical problems from:

painful limbs/breathing problems/heart problems / early death (1 mark for each example)

- (in a heterozygous person) there is a greater chance of passing the faulty gene to children
　　　　　　　　　　　max [8]

3　Similarities:
- both forms of selection involve characteristics that are inherited (i.e. characteristics that are passed on through genes)
- due to changes in genes (mutation) these characteristics are constantly undergoing variation
- and some of these variations will put the organism that inherits them at an advantage in its particular environment over other members of its population.
- these organisms at an advantage survive at the expense of those who have not inherited the particular variation
- the survivors breed and pass on their advantageous characteristic to their offspring
- which in turn, show variation
- the species is thus gradually changing (evolving) over a period of time

Differences:
- in natural selection, it is the organism's environment that is doing the selecting.
- e.g. variation that provides better camouflage for prey increases its chances of avoiding a predator
- and thus increases the chances that this organisms will survive to breed and thus hand on its advantage.
- in artificial selection, it is the human that is selecting
- e.g. selecting the that runs the fastest and using that horse for breeding foals that inherit the ability to run fast　　　max [10]

Chapter 20

1　(a)　• A + above the lake
　　　　• A + above the trees
　　　　• B + above the trees
　　　　• C + close to clouds　　　[4]

(b) • use of fertilisers
   • run off/washed into lake [2]

(c) • it makes up a very high percentage of all cells
   • all chemical reactions in cells/metabolic reactions take place in solution
   • it is an important solvent
   • and transport medium
   • and is used by some animals (e.g. mammals) in temperature regulation max [4]

2 • it takes some time to adjust to the new environment
   • growth rate of the population is slow
   • called the lag phase
   • growth rate increases
   • log/exponential phase
   • food is plentiful
   • rate then slows
   • due to competition (either with its own or other species)
   • (other) limiting factors
   • e.g. food/disease/predators
   • cause the population numbers to remain constant
   • called the stationary phase
   • may lead to possible decline phase max [12]

3 • swellings such as these on the roots of plants are called nodules
   • they occur on the roots of peas and beans, i.e. leguminous plants
   • each nodule contains many bacteria of a type known as 'nitrogen fixing bacteria'
   • these bacteria are able to absorb nitrogen from the air in the soil around them
   • which is used for making their proteins and by the plant for making its proteins
   • when the plant dies, the manufactured protein decomposes
   • first becoming ammonium compounds, then nitrates

• these nitrates are then available for the next crop that the farmer plants
• as they are released relatively slowly as a result of the decomposition
• they do not have the problems associated with the overuse of fertilisers max [8]

Chapter 21

1 • plant waste (a suitable substrate)
  • is placed in a fermenter
  • with yeast
  • and water
  • carbohydrate (in plant material)
  • is used for anaerobic respiration of the yeast,
  • alcohol is produced
  • which is purified, then separated from the contents of fermenter
                                          max [7]

2 (a) i) • for temperature control
         • at the optimum or best level
      ii) • air/oxygen
         • for respiration*
         • of fungus/*Penicillium**
         (*These marks could be awarded in i))
                                          max [4]

(b) • the gene for insulin production
   • is 'cut' or taken from a human chromosome
   • in a cell of a healthy human
   • a plasmid
   • of a bacterium
   • is 'cut'
   • both 'cuts' are performed by restriction enzymes
   • the human gene is inserted into the plasmid
   • the cut ends of the gene and the plasmid have a matching structure **OR** are both described as 'sticky ends'
   • ligase enzymes are the used for joining sticky ends max [8]

3
- blood forms a clot amongst the fibres of the cloth
- the clot is made up of a long-chained fibrous protein (fibrin) that is insoluble
- the red pigment, hemoglobin is bound up within the fibrin
- depending on its contents, a non-biological powder might have difficulty dissolving the insoluble fibrin
- a biological washing powder that contains the enzyme protease
- is able to digest both the fibrin and the hemoglobin (also a protein)
- first into polypeptides, then into amino acids
- amino acids are soluble and thus will wash out of the cloth
- washing at temperatures that are too high (above 60 °C) will denature the protease enzymes
- which will then not work
- lower temperatures (down to 20 °C) will still remove the stain, but will take much longer to do so max [9]

Chapter 22

1 Insecticides:
- washed from the land
- kill insects in rivers/ponds/streams
- which are part of food chains for other animals

chemical waste:
- from industry/factories
- heavy metals/example given (e.g. copper/ mercury)
- kills aquatic organisms
- gets into human food chain

untreated sewage:
- contains pathogens **OR** disease-causing organisms
- which cause diseases such as cholera
- the contaminated water may be used for drinking

- bacteria decompose sewage removing oxygen
- lack of oxygen kills aquatic life max [10]

2 (a) i) greenhouse gases [1]

   ii)
   - they form a blanket around the Earth
   - radiation from the Sun to
   - enters Earth's atmosphere
   - but the gases prevent radiation from leaving
   - the radiation entering is short wave **or** the radiation not able to leave is long wave
   - this increases temperatures of Earth's atmosphere **OR** leads to global warming max [4]

  (b)
  - until 1800 no industrialisation or no machinery
  - CO_2 concentrations remained constant
  - after 1800 – burning of fossil fuels released CO_2
  - deforestation led to less CO_2 absorbed
  - for photosynthesis
  - intensive livestock farming increased methane
  - decomposition in landfill sites produces methane
  - burning fossil fuels produces carbon dioxide
  - both gases increasing + data from table 80 ppm CO_2/1 ppm
  - methane or c. 30% increase for CO_2/c. 130% increase for methane max [6]

3 (a) (i)
   - carbon monoxide is a gas
   - when breathed in, it diffuses through the alveolus walls into the blood
   - once in solution in the blood plasma
   - it is taken up by hemoglobin in the red blood cells
   - once hemoglobin has combined with carbon monoxide (to form carboxyhemoglobin)

- it cannot absorb oxygen
- this results in insufficient oxygen for respiration
- and thus limits the amount of energy released in body cells   max [4]

(ii)
- sulfur dioxide
- when released into the air, immediately dissolves in water vapour in the atmosphere
- when that vapour eventually falls in rain, the rain consequently has a lowered pH
- and is called 'acid rain'
- the effect on plants of acid rain is to damage leaves
- if the damage is severe, it can kill the plant
- lakes that become acidic may not be suitable habitats for the organisms living in them

- if only one link in a food chain is removed, this will affect the balance of life in that lake
- also, acidic water can dissolve harmful chemicals normally locked away in the mud at the bottom once in solution, these harmful chemicals can kill aquatic organisms
- e.g. aluminium ions can be toxic to fish   max [4]

(b) (i)
- methane is a volatile organic compound (VOC)
- so, too, are CFCs – used in refrigerators and aerosol sprays

(ii)
- nitrogen is a constituent of all proteins,
- and thus if organic matter (including fossil fuels) is burnt
- this nitrogen will be oxidised to form oxides of nitrogen   max [2]

Glossary

ACCOMMODATION: The changing of the shape and therefore the focal length of the lens in the eye in order to focus on objects at different distances.

ACID RAIN: A dilute solution of acids that falls to earth when mainly oxides of nitrogen and sulfur in the atmosphere dissolve in rain.

ACTIVE IMMUNITY: Defence against a pathogen as a result of antibody production by the body.

ACTIVE SITE: Section on the surface of an enzyme where a substrate molecule fits exactly and is split into product molecules. The 'lock' in the 'lock and key hypothesis'.

ACTIVE TRANSPORT: An energy-consuming process where substances are transported through living membranes against a concentration gradient.

ADRENAL GLAND: Gland situated above the kidneys. Produces the hormone adrenaline.

ADRENALINE: Hormone produced by the adrenal glands that produces the body's response in times of fear or anger.

AEROBIC RESPIRATION: The release of relatively large amounts of energy by using oxygen to break down foodstuffs. Usually takes the form of the oxidation of glucose in the cytoplasm of living cells.

AIDS: Acquired immune deficiency syndrome. Caused by HIV, a virus that affects the body's ability to fight infection.

ALLELES: A pair of matching genes.

ALVEOLI (SINGULAR: ALVEOLUS): Air sacs of the lungs.

AMINO ACIDS: Simple, soluble units. A few linked together form a polypeptide; many linked together form a protein. Used in cells for building up proteins as the cells grow, and for making special proteins such as enzymes.

AMNION: Membrane that surrounds a developing fetus. It forms the amniotic sac, enclosing the fetus in amniotic fluid.

AMNIOTIC FLUID: A water bath that encloses a developing fetus.

AMYLASE: An enzyme that digests starch to sugars.

ANAEMIA: A lack of hemoglobin, often caused by low levels of iron in a person's diet.

ANAEROBIC RESPIRATION: The release of relatively small amounts of energy by the breakdown of food substances. Occurs in the absence of oxygen.

ANTAGONISTIC MUSCLES: Two muscles that provide opposing forces for movement. One of the pair contracts while, at the same time, the other relaxes.

ANTIBIOTICS: Drugs (e.g. penicillin) used to treat diseases caused by bacteria.

ANTIBODIES: Chemicals, produced by lymphocytes, that 'stick' to bacteria and clump them together, ready for ingestion by phagocytes.

ANTITOXINS: Types of antibodies produced by lymphocytes. They neutralise toxins in the blood.

ARTERY: Vessels with thick, muscular walls that carry blood, under high pressure, away from the heart. A large artery is called an aorta; a small one is called an arteriole.

ARTIFICIAL SELECTION: The deliberate breeding of organisms with particular characteristics, otherwise known as selective breeding.

ASEXUAL REPRODUCTION: The production of genetically identical offspring from one parent.

ATHEROMA: Fatty deposits that form on the walls of arteries, produced by a combination of saturated (animal) fats and cholesterol.

ATRIA (SINGULAR: ATRIUM): Two upper chambers of the heart that receive blood from the body and the lungs.

AUTOTROPHIC: Describes organisms (e.g. plants) able to produce their own food by using small molecules in the environment to build large organic molecules.

AUXIN: A plant hormone that affects the growth of cells.

BACTERIA (SINGULAR: BACTERIUM): Unicellular organisms.

BALANCED DIET: Food and drink consumed by a person which has the correct amount of each constituent (e.g. proteins, carbohydrates) to enable them to be healthy.

BENEDICT'S SOLUTION: Used to test for the presence of certain sugars including maltose and glucose (i.e. 'reducing' sugars).

BINOMIAL SYSTEM: The use of a two-word (usually) Latin name for a species of organism.

BIRTH CONTROL: Methods of preventing pregnancy (e.g. through the use of contraceptives).

BIURET SOLUTIONS: Used to test for the presence of proteins.

BLASTOCYST: A stage in embryonic development after the zygote has divided to form a hollow ball of cells.

BLIND SPOT: The point in the eye where the retina is joined to the optic nerve. There are no rods or cones, so images formed here are not converted into impulses and relayed to the brain.

CAPILLARIES: Microscopic blood vessels that carry blood from arterioles to venules.

CAPILLARITY: The movement of liquids upward through very narrow tubes.

CARBOHYDRATE: Organic chemicals containing only the elements carbon, hydrogen and oxygen. Ratio of hydrogen atoms to oxygen atoms is always 2:1.

CARNIVORES: All consumers above the level of herbivore, i.e. all meat eaters.

CARPELS: The female parts of a flower – a stigma, connected by a style to the ovary, in which lie the ovules which contain the female gamete.

CATALYSTS: Particular chemicals that can affect how quickly chemical reactions occur (usually speed up reactions).

CELL MEMBRANE: Outer covering of the cell that controls the passage of substances into and out of the cell.

CELL WALL: A 'box' made of cellulose that encloses the plant cell – not present around animal cells.

CENTRAL NERVOUS SYSTEM (CNS): The body's coordinating centre, made up of the brain and the spinal cord. Receives information about the environment from receptors and directs a response to effectors (muscles or glands).

CHLOROPHYLL: A green pigment found within the chloroplasts of plant cells. Traps sunlight for use in the process of photosynthesis. Contains magnesium.

CHLOROPLASTS: Small bodies lying in the cytoplasm of those plant cells involved in photosynthesis. Green in colour because they contain chlorophyll.

CHROMOSOME: Possesses genes which are responsible for programming the cytoplasm to manufacture particular proteins.

CLONES: A population of organisms produced by asexual reproduction, and all genetically identical.

CLOT: A clump of blood cells trapped in a mesh of fibrin. It prevents the entry of bacteria at a wound. On the skin surface it dries and hardens to form a scab.

CNS: Central nervous system.

CODOMINANCE: A type of monohybrid inheritance, when both alleles have an equal effect on the phenotype of the offspring.

CONCENTRATION GRADIENT: When a region of (relatively) high concentration of molecules or particles is next to a region of (relatively) low concentration. Must be present for diffusion to occur.

CONES: Light-sensitive cells in the retina of the eye that provide a picture with greater detail and in colour. They convert light energy into electrical energy.

CONSUMER: Any organism which relies on the energy supplied by the producer in its food chain.

CONTINUOUS VARIATION: Where both inherited and environmental factors determine the characteristics of an individual (e.g. body mass, height).

CONTROL: Apparatus and materials identical to those in an experiment but lacking in the one feature being investigated. Used to make a comparison with the experiment in order to make the results of the experiment valid.

COPD: (Chronic obstructive pulmonary disease) The name given to a collection of conditions that affect the lungs (including chronic bronchitis and emphysema) causing difficulty in breathing.

CORNEA: Transparent part of the eye that allows light rays to enter and refracts them towards each other.

CORONARY ARTERY: Vessel that supplies oxygenated blood to the heart muscle.

COTYLEDONS: Organs in seeds, often used for storing starch and protein.

CYTOPLASM: A jelly-like substance in which the chemical reactions of the cell take place.

DCPIP: A chemical compound used for measuring the vitamin C content of a sample.

DEAMINATION: Process by which excess amino acids are broken down in the liver to produce the excretory chemical urea.

DECOMPOSERS: Organisms that release enzymes to break down large molecules in dead organic matter into smaller ones that can then be recycled.

DEFICIENCY DISEASE: Conditions caused by a lack of a constituent (e.g. vitamin C, vitamin D, calcium or iron) in a person's diet.

DERMIS: Lower layer of skin containing most of the skin structures (e.g. sweat glands, venules, arterioles).

DETOXIFICATION: The removal and breakdown of toxins (e.g. alcohol) from the blood. A major function of the liver.

DICHOTOMOUS KEY: A series of questions used to identify an organism. Each question is applied to the organism in turn, and has two possible answers (a dichotomy is a division into two). Depending on the answers that apply, so the organism can be identified.

DIFFUSION: The movement of molecules from a region of higher concentration to a region of lower concentration, down a concentration gradient.

DISCONTINUOUS VARIATION: Where inheritance alone determines the characteristics of an individual (e.g. blood group).

DISSOLVE: Mix a substance into a liquid so that it is absorbed into the liquid.

DRUGS: Externally administered substances which modify or affect chemical reactions in the body.

ECG: (Electrocardiogram) A record of the electrical activity of the heart.

ECOSYSTEM: A community of organisms living together in a habitat and connected through food webs.

ENDOPLASMIC RETICULUM: A network of membranes that runs throughout the cytoplasm. When ribosomes are attached to the membranes they form 'rough endoplasmic reticulum'.

ENZYMES: Biological catalysts that control chemical reactions in living organisms. Each has a specific shape and works most effectively at a particular temperature and pH.

ETHANOL: Alcohol used to test for the presence of fats. A waste product of anaerobic respiration in yeast.

EUTROPHICATION: The abundant growth of water plants. Accelerated when nitrate levels increase in waterways.

EVOLUTION: Gradual change in the characters of a species through natural selection. Takes place over many generations.

EXCRETION: The removal of waste products of metabolism from organisms.

EXPIRATION: The breathing out of air into the atmosphere.

EXTERNAL DIGESTION: Method of nutrition, characteristic of saprotrophs, by the release of enzymes onto an organic substrate ('food').

FATS: Insoluble organic molecules containing the elements carbon, hydrogen and oxygen only. Ratio of hydrogen to oxygen is much higher than 2:1. Formed by the joining of a glycerol molecule with fatty acid molecules.

FATTY ACIDS: Soluble molecules that, when joined with glycerol, form fat.

FERMENTATION: The anaerobic decomposition of some organic substances (e.g. of sugar to alcohol). Carbon dioxide is a waste product.

FETUS: A developing embryo in its mother's uterus.

FIBRIN: An insoluble, stringy protein formed by fibrinogen and enzymes released by damaged cells. It forms a mesh that traps blood cells and becomes a clot.

FIBRINOGEN: A soluble protein found in blood that plays a part in blood clotting.

FITNESS: A measure of how likely it is that an organism will survive and reproduce in its environment.

FLACCID: Used to describe cells, tissues or organs when they lose their shape and firmness (turgor).

FOLLICLE STIMULATING HORMONE (FSH): A hormone released by the pituitary gland that stimulates the ovaries to produce and release ova (eggs).

FOOD CHAIN: A sequence of organisms, starting with a photosynthesising organism (usually a green plant), through which energy is passed as one organism is eaten by the next in the sequence.

FOOD WEB: Interlinked food chains involving organisms within the same ecosystem.

FOVEA: A very sensitive part of the retina that has far more cones than rods. Also called the yellow spot.

FUNGI (SINGULAR: FUNGUS): Parasitic or saprotrophic multicellular organisms that feed on organic matter by digesting and absorbing it. They do not photosynthesise.

GAMETES: Male or female sex cells.

GASEOUS EXCHANGE: The simultaneous absorption and release of gases by an organism. For example, mesophyll cells in plants absorb carbon dioxide and release oxygen during photosynthesis; cells of the alveoli pass oxygen from the lungs into the blood and carbon dioxide in the opposite direction.

GENE: A unit of inheritance, part of a chromosome.

GENETIC ENGINEERING: Artificially changing the genetic make-up of cells.

GENOTYPE: Genetic combination of an individual. Three possibilities are: homozygous dominant, homozygous recessive, heterozygous.

GLYCEROL: Molecule that, when joined with fatty acids, forms fat.

GLYCOGEN: A carbohydrate with large, insoluble molecules. It is stored in the cells of the liver and muscles and in fungal cells. The conversion of glucose to glycogen takes place in the liver of mammals. This process is controlled by the hormone insulin, secreted by the pancreas.

GONADS: The organs that produce gametes (reproductive cells): the testes in males; ovaries in females.

GRAVITROPISM: The growth response of a plant organ to the stimulus of gravity.

HEMOGLOBIN: Iron-containing pigment found in the cytoplasm of red blood cells. It carries oxygen around the body by combining with it in the lungs to become oxyhemoglobin.

HERBIVORES: Consumers that feed directly on the producer in their food chain.

HETEROZYGOUS: Having a pair of dissimilar alleles for a particular character.

HIGH WATER POTENTIAL: Dilute solutions with a relatively large number of water molecules.

HIV: Human immunodeficiency virus.

HOMEOSTASIS: The maintenance of a constant internal environment in the body. Performed by organs of homeostasis (e.g. the skin).

HOMOZYGOUS: Having a pair of similar alleles for a particular character (e.g. both dominant, or both recessive).

HORMONE: A chemical substance, produced by a gland and carried by the blood, which alters the activity of one or more specific target organs. It is then destroyed in the liver.

HUMUS: Formed when dead organic matter decomposes in the soil. Humus provides a steady supply of ions. It acts as a sponge, soaking up and holding water in the soil, and helps to bind the soil together, preventing soil erosion.

HYDROLYSIS: Enzyme-controlled chemical reaction that involves the introduction of a water molecule in order to split a substrate molecule. The newly exposed ends of product molecules are 'sealed' so they will not rejoin after being split, common in digestion.

HYPOTHALAMUS: The part of the brain responsible for monitoring changes in the blood.

INSOLUBLE: Unable to be mixed into and absorbed by a liquid (dissolved).

INSPIRATION: The taking in, or breathing in, of air from the atmosphere.

INSULIN: Hormone produced by the islets of Langerhans in the pancreas, involved in the uptake of glucose by cells and its conversion into glycogen.

IODINE SOLUTION: Used to show the presence of starch (by turning blue/black), also as a temporary stain for plant cells.

ION: A charged atom or group of atoms formed when a molecule dissolves in water (e.g. potassium nitrate dissolves to form potassium+ and nitrate− ions).

IRIS: Part of the eye that controls the intensity of light falling on the retina. It has an antagonistic arrangement of circular and radial muscles.

IRON DEFICIENCY: Low levels of iron in a person's diet. Leads to a lack of hemoglobin, which is necessary for carrying oxygen around the body.

IUD/IUS: (Intra-uterine device or system) Forms of contraceptive inserted into the uterus to prevent implantation of a fetus. An IUS works by slowly releasing chemicals.

KIDNEY DIALYSIS: The use of a machine to perform the functions of a kidney. It removes chemicals with small molecules (urea, toxins and ions) from blood but does not allow larger molecules (e.g. plasma proteins) to leave.

LENS: Transparent, elastic part of the eye responsible for focusing an image on the retina.

LIGNIN: Chemical that helps to strengthen the walls of xylem vessels in plants.

LIMITING FACTORS: Particular factors that limit the rate of photosynthesis in plants, even when all other factors may be optimum. Examples are light, carbon dioxide, water and temperature.

LIPASE: An enzyme that digests fats to fatty acids and glycerol.

LOW WATER POTENTIAL: Concentrated solutions with fewer water molecules.

LUTEINISING HORMONE (LH): A hormone released by the pituitary gland that stimulates the ovaries to produce ova. After the release of an ovum, it then stimulates the follicle cells in the ovary to produce the hormone progesterone. (Follicle cells were the cells that nourished the ovum during its development in the ovary.)

LYMPHATIC SYSTEM: A system of vessels for returning lymph (tissue fluid plus fats absorbed by the lacteals of the villi) to the blood system.

LYMPHOCYTES: A type of white blood cell, made in the lymph glands. They produce antitoxins and other antibodies that 'stick' to bacteria and clump them together for ingestion by phagocytes.

MAGNESIUM IONS: A form of magnesium absorbed by plants from the soil through the root hair.

MEIOSIS: (Also known as 'reduction division') It is a form of cell division that occurs during gamete production, in which the chromosome number is halved (is changed from the diploid number to the haploid number). It also produces gametes that are all genetically unique.

MENOPAUSE: When a female stops ovulating and can no longer become pregnant. Usually occurs at around 50 years of age.

MENSTRUATION: Stage in the menstrual cycle when blood and the lining of the uterus are passed out of the vagina and vulva.

MESOPHYLL CELLS: Palisade and spongy cells in a leaf, involved in photosynthesis.

METABOLISM: All the chemical reactions occurring in cells.

METAMORPHOSIS: The change in form and feeding habits when a larva develops into an adult.

MICROORGANISMS: Organisms so small that they can be studied only by using a microscope (e.g. viruses, bacteria and some fungi).

MICROVILLI: Microscopic extensions of the cell membrane to increase the surface area of the cell.

MILK TEETH: A person's first set of teeth that last for around 5–4 years, then are pushed out by the permanent teeth.

MITOCHONDRIA: Organelles in the cytoplasm that release energy during aerobic respiration.

MITOSIS: A process of cell division when each chromosome forms an exact replica of itself. The two cells formed are identical to each other and to the original cell.

MOLECULAR COHESION: The tendency for molecules to attract one another and thus stick together – particularly the case with water molecules.

MONOHYBRID INHERITANCE: Inheritance involving only one pair of contrasting alleles.

MORPHOLOGY: The overall shape and form of the body of an organism.

MRSA: A species of pathogenic bacterium that has developed high resistance to antibiotics.

MULTICELLULAR: Living organisms that have many cells.

MUTATION: A spontaneous change in the structure of a gene or chromosome.

NATURAL SELECTION: The survival of those organisms most effectively adapted to their environment.

NEGATIVE FEEDBACK: A system which automatically brings about a correction in the body's internal environment (e.g. temperature), regardless of which side of the optimum the change has occurred.

NEURONES: Individual nerve cells with their own cytoplasm, cell membrane and nucleus.

NEUROTRANSMITTER: The chemical released by a nerve ending when an impulse arrives and that, after passing across a synapse, is responsible for generating an impulse in the next neurone.

NITRATE ION: A form of nitrogen that plants absorb from the soil through root hairs.

NITROGEN FIXATION: Conversion of atmospheric (gaseous) nitrogen into nitrogen compounds that can be used by living organisms.

NUCLEUS: The part of the cell that controls its growth and development. It contains a number of chromosomes made of the chemical DNA.

OESTROGEN: A hormone released by the ovaries that is responsible for the development and maintenance of the female secondary sexual characteristics.

OPTIMUM: The best; particularly refers to a state (e.g. temperature, pH level) when processes can take place most efficiently.

ORGAN: Several tissues working together to produce a particular function.

ORGAN SYSTEM: A collection of different organs working together to perform a particular function.

ORGANISM: A collection of organ systems working together.

OSMOREGULATION: The maintenance of a constant concentration (e.g. of blood plasma, performed by the kidneys).

OSMOSIS: The passage of water molecules from a region of high water potential, to a region of lower water potential, through a partially permeable membrane.

OVA (SINGULAR: 'OVUM'): Female gametes produced in the ovaries.

OXYHEMOGLOBIN: Constituent of red blood cells, formed by the combination of oxygen and hemoglobin.

PARASITE: An organism that obtains its food from another, usually larger living organism ('host'), the host always suffering in the relationship.

PASSIVE IMMUNITY: Immunity acquired through antibodies form another organism – such as in milk from mother to infant.

PATHOGEN: A disease-causing organism (e.g. virus, bacterium).

PENIS: Male organ for introducing sperms into the female.

PERISTALSIS: Waves of muscle contractions. Occurs in the oesophagus (pushing boli towards the stomach), and through the duodenum, ileum and colon (pushing food towards the rectum).

PHAGOCYTE: A type of white blood cell, made in the bone marrow. It has a lobed nucleus and is capable of movement. Its function is to ingest bacteria.

PHAGOCYTOSIS: The ingestion of potentially harmful bacteria by phagocytes. Prevents or helps to overcome infection.

PHENOTYPE: Inherited feature in an individual's appearance.

PHLOEM: Tissue for transporting sugars and amino acids within a plant.

PHOTOSYNTHESIS: A process performed by green plants, in which light energy is converted into chemical energy.

PHOTOTROPISM: The growth response of a plant organ to the stimulus of one-sided light.

PLACENTA: A special structure that carries out the exchange between the mother and fetus of the chemicals involved in the fetus's nutrition, respiration and excretion.

PLASMA: Watery component of blood that carries dissolved chemicals, blood cells and heat.

PLASMID: A separate strand of DNA, often circular in shape, that occurs naturally in bacteria.

PLASMOLYSIS: When a cell's cytoplasm is pulled away from the cell wall as a result of osmosis. Occurs when the cell is placed in a solution of lower water potential, and water is drawn from the vacuole.

PLATELETS: Fragments of cells made in the bone marrow. They play a part in blood clotting and help to block holes in damaged capillary walls.

POLLEN TUBE: Structure produced by a germinating pollen grain. The pollen tube grows down the style towards the ovary by releasing enzymes at its tip to digest the cells of the style beneath.

POLLINATION: The transfer of pollen from an anther to a stigma.

POLYPEPTIDES: Formed by a few amino acids linked together. Enzymes in the body break down proteins to polypeptides, and polypeptides to amino acids.

PRIMARY CONSUMER: (also 'herbivore') A consumer that feeds directly on the producer in its food chain.

PRODUCERS: Organisms which manufacture and supply energy-rich foods, made by photosynthesis, to all organisms in their food chain.

PRODUCT: The molecules produced as a result of enzyme action on substrate molecules.

PROGESTERONE: A hormone produced by the follicle cells, after the release of an ovum, which maintains the spongy lining of the uterus and stops the pituitary gland producing FSH.

PROTEASE: An enzyme that digests proteins to amino acids.

PROTEINS: Contain the elements carbon, hydrogen, oxygen and nitrogen. Often contain other elements such as sulfur and phosphorus. Built up from amino acids.

PROTOPLASM: The cytoplasm and the nucleus of a cell.

PUBERTY: A stage in life when the release of hormones activates the reproductive organs. In humans, this occurs around the age of 12 years.

PYRAMID OF BIOMASS: A diagram constructed using the dry mass of organisms at each trophic level in a food web, with the producer at the base of the pyramid and the top consumer at the apex (top).

PYRAMID OF NUMBERS: A pyramid-shaped diagram composed of blocks similar to a pyramid of biomass but, this time, the width of each block indicates the number of organisms at each trophic level.

RBC: Red blood cell.

RECESSIVE GENES: Characters determined by these genes will not appear in an individual unless two recessive (i.e. no dominant) alleles are present.

RED BLOOD CELLS: Small, biconcave and flexible cells that carry oxygen around the body.

REDUCING SUGARS: A group of sugars, including maltose and glucose, that, when reacting with Benedict's solution, act as chemicals known as 'reducing agents'.

REFLEX: A fast, coordinated, automatic response to a specific stimulus.

RENAL CAPSULE: The cup-like structure at the beginning of each kidney tubule that contains a knot of blood capillaries (the glomerulus) and is responsible for the high-pressure filtration of blood.

RESPIRATION: The release of energy from food substances, that takes place in all living cells to perform all their functions.

RETINA: The innermost, light-sensitive layer of the eye.

RIBOSOME: One of many minute bodies that lie attached to membranes in the cytoplasm. They are where amino acids are linked together to make proteins.

RICKETS: A deficiency disease of bones caused by a lack of vitamin D in a person's diet.

RODS: Light-sensitive cells found in the retina of the eye. Important when light intensity is low. They convert light energy into electrical energy.

ROOT HAIR CELL: Plant cell specially adapted to absorb water and mineral ions (salts) from the soil.

ROOT PRESSURE: Created by the process of osmosis carrying water across the root to the vascular bundle of the stem. The pressure forces water into the xylem and pushes it along the root towards the stem.

SAPROTROPH: Organism that feeds on dead organic matter through external digestion.

SCAB: A dried and hardened clot that covers a wound until the skin beneath has repaired.

SCURVY: A deficiency disease of the gums and skin caused by a lack of vitamin C in a person's diet.

SECONDARY CONSUMER: A consumer which feeds directly on the herbivore in its food chain.

SEMEN: Sperms and seminal fluid.

SEMINAL FLUID: A nutrient fluid in which sperms are able to swim.

SEX CHROMOSOMES: One pair of chromosomes that determine the sex of an offspring.

SEXUAL REPRODUCTION: The fusion of male and female nuclei to form a zygote. Zygotes develop into offspring genetically different from each other and from their parents.

SPECIES: A group of organisms that interbreed to produce fertile offspring.

SPERM: An abbreviation for 'spermatozoon' – the male gamete (or sex cell).

SPIRACLE: A breathing hole in the side of a thoracic or abdominal segment of an insect.

STAMEN: The anther and filament of a flower, involved in the production of pollen grains.

STARCH: Insoluble carbohydrate produced by plant cells from glucose. Stored in the chloroplasts of photosynthesising cells and many storage organs of plants.

STOMATA (SINGULAR: STOMA): Pores through which gases diffuse into and out of a leaf.

SUBSTRATE: Molecule on which a catalyst works, changing it into product molecules. The 'key' in the 'lock and key hypothesis'. Also, the food on which organisms such as bacteria and fungi grow.

SYNAPSE: The gap between the dendrites (nerve endings) of neighbouring neurones.

TEMPERATURE REGULATION: Function performed by the skin to maintain body temperature, in humans, at 37 °C. Includes sweating, dilation or constriction of arterioles and the control of blood flow to the skin.

TERTIARY CONSUMER: A consumer that feeds directly on the secondary consumer in its food chain.

TESTOSTERONE: A hormone produced by the testes in males from the age of puberty. It controls the development and maintenance of secondary sexual characteristics.

TEST CROSS: A genetic cross involving one homozygous recessive parent.

TISSUE: Many similar cells working together and performing the same function.

TISSUE CULTURE: Commercial application of asexual reproduction in plants, in which pieces of tissue are removed from an organism and grown in an artificial medium in sterile conditions.

TISSUE FLUID: Blood without red blood cells, plasma proteins and some white blood cells. It bathes the body's cells.

TISSUE REJECTION: When the body's immune system fails to accept a transplanted organ (e.g. heart or kidney) and attempts to destroy it as a harmful protein.

TOLERANCE: The ability of an organism to take progressively increased dosages of a drug.

TOXINS: Poisons.

TRANSLOCATION: The movement of chemicals around a plant.

TRANSPIRATION: The evaporation of water from the mesophyll cells of a leaf and the removal of that vapour through the stomata of the leaf.

TRANSPIRATION PULL: A force created by transpiration, where water is drawn up to the leaf to replace the water that has been lost.

TRANSPIRATION STREAM: A continuous stream of water and ions that travels up a plant.

TURGOR: The pressure created as water enters a plant cell, causing the cytoplasmic lining of the cell to press against the cell wall. Helps to make plant cells firm (turgid).

UMBILICAL CORD: Connection between placenta and fetus. The fetal blood vessels run within it.

UNICELLULAR: Made up of one cell only, as in the simplest living organisms.

UREA: A nitrogenous waste product that passes in the blood from the liver to the kidneys for excretion in urine.

URINE: Waste solution containing urea, ions, toxins and water.

VACUOLE: A large, central space in plant cells that contains cell sap, a solution made up mostly of sugars. Also called the 'sap vacuole'.

VACCINATION: The exposing of an organism to material (an antigen) that stimulates the organism's white blood cells to produce antibodies to destroy the antigen.

VASCULAR BUNDLES: Contain the tissues for transport within a plant – xylem (for carrying water and ions) and phloem (for carrying sugar and amino acids).

VEIN: Vessel which carries blood under low pressure towards the heart – thinner walls than arteries. A large vein is a vena cava; a small one is a venule.

VENTRICLES: Two lower chambers of the heart that pump blood out of the heart.

VILLI: Microscopic finger-like projections found on the walls of the ileum. Designed to maximize surface area to allow food absorption.

VIRUSES: Very small, parasitic organisms that cause disease. They cannot be treated with antibiotics. They do not possess all the characteristics of living organisms.

WILT: When a plant loses its rigidity as a result of a loss of turgidity in its cells. It occurs when the transpiration rate exceeds the rate that water can be absorbed from the soil and water starts to be lost from the plant's cells.

WITHDRAWAL SYMPTOMS: Unpleasant effects that result when someone stops taking a drug to which they are addicted (e.g. heroin).

XYLEM VESSELS: Tube-like structures in plants specially adapted to conduct water and ions from the roots to the stem, leaves, flowers and fruits. Also provide support for the parts of the plant above the ground.

ZYGOTE: Formed by the fusion of male and female nuclei during sexual reproduction. Zygotes develop into offspring genetically different from each other and from their parents.

Index

Index